Subsurface Geologic Investigations of New York Finger Lakes: Implications for Late Quaternary Deglaciation and Environmental Change

Edited by

Henry T. Mullins
Department of Earth Sciences
Heroy Geology Laboratory
Syracuse University
Syracuse, New York 13244

and

Nicholas Eyles
Department of Environmental Sciences
University of Toronto
Scarborough Campus
Scarborough, Ontario M1C 1A4, Canada

311

1996

Published by The Geological Society of America, Inc.
3300 Penrose Place, P.O. Box 9140, Boulder, Colorado 80301

Printed in U.S.A.

GSA Books Science Editor Abhijit Basu

Library of Congress Cataloging-in-Publication Data
Subsurface geologic investigations of New York Finger Lakes:
implications for Late Quaternary deglaciation and environmental
change / edited by Henry T. Mullins and Nicholas Eyles.
p. cm. -- (Special paper ; 311)
Includes bibliographical references and index.
ISBN 0-8137-2311-6
1. Glacial epoch--New York (State)--Finger Lakes Region.
2. Geology--New York (State)--Finger Lakes Region.
3. Paleoclimatology--Holocene. 4. Paleoclimatology--New York
(State)--Finger Lakes Region. I. Mullins, Henry T. II. Eyles, N.
III. Series: Special papers (Geological Society of America) ; 311.
QE697.S898 1996
551.7'92'09747--dc20 96-35721
CIP

Cover: Satellite photograph of New York Finger Lakes taken March 23, 1973 (ERTS-1243-15244-5). Large lake in upper left-hand corner is Lake Ontario. Data available from U.S. Geological Survey, EROS Data Center, Sioux Falls, South Dakota.

10 9 8 7 6 5 4 3 2 1

Contents

Preface

The Finger Lakes of New York State, which derive their name from Native American legend of the Great Spirit placing his hands down upon the Earth, have long been considered one of the world's great testimonies to continental glaciation. This region of 11 elongate lakes has attracted the attention of glacial geologists and geomorphologists ever since Agassiz spoke of its glacial "heritage" in an address at Cornell University in 1868. However, despite scientific inquiry spanning more than a century, the origin and evolution of these world-renowned lakes have remained largely unresolved.

Perhaps the most significant limitation on our knowledge and understanding of the Finger Lakes has been a paucity of data from beneath the lakes. Onland geologic investigations during the past century recently culminated in the publication (1986) of a marvelous surficial geologic map of the Finger Lakes region, yet the subsurface geology of the region, particularly that in the water-covered areas, is still poorly known. Although drill records from the turn of the century hinted at deep-scour and thick sediment-infill of the Finger Lakes basins, it was not until the 1960s that the first seismic reflection records from Seneca Lake confirmed these earlier reports.

In 1984, drawing on prior marine geophysical experience, we conducted a preliminary seismic reflection survey of northern Seneca Lake with the help of a "starter grant" from Syracuse University. These data formed the basis of a two-phase research grant funded by the National Science Foundation, which had three primary scientific objectives: (1) the provincial question of the origin and evolution of the Finger Lakes; (2) regional inquiry into the tempo and mode of Laurentide ice sheet deglaciation; and (3) the more global concern of postglacial climatic and environmental change. During phase I (1986–1989), we expanded our seismic reflection data base from Seneca Lake to all 11 Finger Lakes, working from the 6-m *R/V Alexander Winchell.* This resulted in the acquisition of ~1,300 km of high-resolution seismic reflection profiles from the Finger Lakes and ~20 km of reflection data from adjacent onland areas.

Based on these geophysical data, phase II (1990–1993) of the Finger Lakes research project involved the recovery of a 120-m-long drillcore, and downhole geophysical data, from the dry lake valley south of Canandaigua Lake. These drill results provided the necessary stratigraphic record for our geophysical data to address our first two scientific objectives and to obtain samples for paleoenvironmental analysis. Combined, the geophysical and drill data have defined a thick sediment record of Laurentide ice sheet deglaciation and postglacial environmental change.

The purpose of this volume is to comprehensively report results from phase I (reflection profiling) and phase II (drillcoring) of the Finger Lakes research project and to discuss their implications for deglaciation and postglacial environmental changes. Much of the research reported here was conducted as student dissertations, theses, and independent studies at Syracuse University. I thank my co-editor, Nick Eyles, University of Toronto, for providing an objective editorial perspective. I also acknowledge outside reviewers of the chapters presented here: Parker Calkin, State University of New York, Buffalo; Paul Carlson, U.S. Geological Survey, Menlo Park; Peter Clark, Oregon State University; Walter Dean, U.S. Geological Survey, Denver; Steve Forman, Ohio State University; Roger Hooke, University of Minnesota; Peter Knuepfer, State University of New York, Binghamton; Doug Nelson, Syracuse University; and Robert Sheridan, Rutgers University. I also thank the Smith brothers of Naples, New York, for allowing us permission to drill on their land. And, finally, I am grateful for the unending patience of former GSA Books Science Editor Richard Hoppin, University of Iowa.

I hope that the reader will find this collection of papers an informative and intriguing look at the Quaternary subsurface geology of the Finger Lakes.

Henry T. Mullins
Editor
October 2, 1995

Geological Society of America
Special Paper 311
1996

Seismic stratigraphy of the Finger Lakes: A continental record of Heinrich event H-1 and Laurentide ice sheet instability

Henry T. Mullins
Department of Earth Sciences, Heroy Geology Laboratory, Syracuse University, Syracuse, New York 13244
Edward J. Hinchey
ERM-Northeast, 5788 Widewaters Parkway, Dewitt, New York 13214
Robert W. Wellner*
Department of Geology, University of Alabama, Tuscaloosa, Alabama 35487-0338
David B. Stephens
Worcester Academy, 81 Providence Street, Worcester, Massachusetts 01604
William T. Anderson, Jr.,* and Thomas R. Dwyer*
Department of Earth Sciences, Heroy Geology Laboratory, Syracuse University, Syracuse, New York 13244
Albert C. Hine
Department of Marine Sciences, University of South Florida, St. Petersburg, Florida 33701

ABSTRACT

Seismic reflection surveys of 8 of the 11 Finger Lakes of central New York State have documented the deep (as much as 306 m below sea level) glacial scour of these lake basins and their subsequent infill by thick (up to 270 m) unconsolidated sediment. Drill data indicate that sediment infill occurred rapidly during a short interval between ~14,400 and 13,900 14C yr ago, coeval with Heinrich event H-1 when large volumes of icebergs and meltwater were discharged into the North Atlantic during an unstable phase of the Laurentide ice sheet.

Six acoustically defined depositional sequences beneath the lakes, correlated with drillcore and piston core samples, record the infill history of the Finger Lakes during the late Wisconsin. Depositional sequence I is equivalent to thick, water-laid sands and gravels of the Valley Heads moraine deposited ~14.4 ka. During retreat of the ice margin from its Valley Heads position, subglacial meltwaters transported large volumes of fine-grained sediment into the Finger Lake basins (sequences II and III). Sequence IV records a phase of high-level proglacial lakes when ice blocked northern outlets of the Finger Lakes and fine-grained sediments continued to be transported into the basins from the north. An abrupt drop of proglacial lake levels and a drainage reversal is recorded by sequence V when sediments first began to enter the Finger Lakes from the south following retreat of the ice margin past the northern outlets of the lakes. The well-known modern glens and waterfalls of the Finger Lakes region formed at this time when lateral streams adjusted to dramatically lowered

*Present addresses: Wellner, Exxon Production Research Company, P.O. Box 2189, Houston, Texas 77252-2189; Anderson, Geological Institute, Swiss Federal Institute of Technology (ETH), Zurich, Switzerland; Dwyer, Blasland, Bouck and Lee, Inc., 6723 Towpath Road, Syracuse, New York 13214.

Mullins, H. T., Hinchey, E. J., Wellner, R. W., Stephens, D. B., Anderson, W. T., Jr., Dwyer, T. R., and Hine, A. C., 1996, Seismic stratigraphy of the Finger Lakes: A continental record of Heinrich event H-1 and Laurentide ice sheet instability, *in* Mullins, H. T., and Eyles, N., eds., Subsurface Geologic Investigations of New York Finger Lakes: Implications for Late Quaternary Deglaciation and Environmental Change: Boulder, Colorado, Geological Society of America Special Paper 311.

local base levels. Radiocarbon dates across the sequence V/VI boundary indicate that proglacial lake levels dropped ~13.9 ka. Depositional sequence VI represents thin (<15 m) postglacial sediments that have accumulated in the Finger Lakes following the establishment of "modern" drainage.

Definition of deep scour and a rapid phase of sediment infill beneath the Finger Lakes 14.4 to 13.9 ka not only provide a continental record of Heinrich event H-1 but also support models of rapid ice flow and large-scale instability of the Laurentide ice sheet at this time.

INTRODUCTION

The stability of continental ice sheets is currently of great interest because of the impact ice sheets have had on past global climates as well as their potential impact for future global change (Hughes, 1992; Wright et al., 1993; Broecker, 1994). Ice sheets are also critical links, both direct and indirect, in the behavior of the global hydrosphere-atmosphere system (Clark, 1994; Blanchon and Shaw, 1995).

Evidence for general ice sheet instability, surges, and collapse (Hughes, 1987) has come from the study of both modern and ancient ice sheets. The rapid streaming of ice within the Antarctic ice sheet is now well established (Bentley, 1987; Clarke, 1987) and it has long been known that the terrestrial component of the Laurentide ice sheet experienced unstable behavior of its margin during retreat from the last glacial maximum (Clark, 1994). Further evidence for instability of the Laurentide ice sheet has recently come from marine sediments in the north Atlantic. A series of at least six layers of ice-rafted debris, rich in detrital carbonate, have been identified in deep-sea sediments spanning the past 70,000 yr (Heinrich, 1988; Bond et al., 1992; Broecker et al., 1992; Andrews and Tedesco, 1992; Grousset et al., 1993; Andrews et al., 1994; Broecker, 1994). These ice-rafted layers or Heinrich events, imply discrete episodes of massive, but short-lived, outpourings of icebergs from ice streams during phases of ice sheet instability, and perhaps global climate change.

Unlike the marine record, the terrestrial record of continental ice sheet dynamics is notorious for being relatively thin, incomplete, and discontinuous. This is particularly acute in New York State where onland morphostratigraphic units are discontinuous and difficult to trace as well as date (Muller and Calkin, 1993). In contrast, glacial lake basins commonly contain thick, well-preserved records of deglaciation (Mullins et al., 1990, 1991a; Eyles et al., 1990, 1991). Mullins and Hinchey (1989) previously reported on the thick (up to 275 m), continuous nature of the late Quaternary (<~14 ka)[1] sediment fill beneath the Finger Lakes. They suggested that the Finger Lakes might provide a critical testing ground for deglaciation models of the southern margin of the Laurentide ice sheet because of their thick, well-preserved sediment record.

This chapter expands on the first-order data of Mullins and Hinchey (1989) by presenting detailed results of a comprehensive seismic reflection investigation of the Finger Lakes. We also discuss correlation of the Finger Lakes stratigraphic record with Heinrich event H-1 and its general implication for deglaciation of the Laurentide ice sheet.

REGIONAL SETTING

The Finger Lakes of central New York State consist of 11 elongate, glacially scoured lake basins (von Engeln, 1961). Located south of Lake Ontario (Fig. 1) along the northern margin of the glaciated Appalachian Plateau (Coates, 1968, 1974), the Finger Lakes have been eroded into undeformed, but well-jointed, Devonian sedimentary rocks (largely shale) that dip gently to the south-southwest. The seven larger, eastern Finger Lakes (Otisco, Skaneateles, Owasco, Cayuga, Seneca, Keuka, Canandaigua) form a radiating pattern that projects northward into the eastern basin of Lake Ontario, whereas the four smaller, western Finger Lakes (Honeoye, Canadice, Hemlock, Conesus) project northward to a point near the city of Rochester (Fig. 1). The lakes vary considerably in size, ranging in length from 5 to 61 km, in lake-level elevation from 116 to 334 m, and in maximum water depth from 9 to 186 m (Table 1).

North of the Finger Lakes (Fig. 2), is the Ontario Lowland characterized by an extensive drumlin field (White, 1985; Ridky and Bindschadler, 1990) and a complex system of meltwater channels (Muller and Cadwell, 1986), including Montezuma wetlands north of Cayuga Lake (see Petruccione et al., this volume). The uplands between the Finger Lakes are covered by a thin layer of till with a series of distinct chevron-shaped till moraines (Fig. 2), which become more laterally continuous to the north (Muller and Cadwell, 1986; Muller and Calkin, 1993).

Immediately south of the Finger Lake basins, and restricted to the valleys, are kame moraines (Fig. 2) (Muller and Cadwell, 1986) collectively referred to as the Valley Heads Moraine. The Valley Heads kame moraines are thick (locally >200 m; see Wellner et al., this volume) accumulations of largely coarse-grained, water-laid drift. Based on morphostratigraphic correlations, Fullerton (1986) suggested an age of 12.95 to 14.1 ka for the Valley Heads. Krall (1977) inferred an age of 14.8 ka for the Cassville-Cooperstown moraine that is truncated by the Valley Heads, indicating that Valley Heads must be less than 14.8 ka. Wellner et al. (this volume) have also reported a radiocarbon age of 13,650 ± 210 yr from an in situ peat layer that is stratigraphically above the Valley Heads in

[1]All ages reported in this chapter are in uncalibrated radiocarbon years.

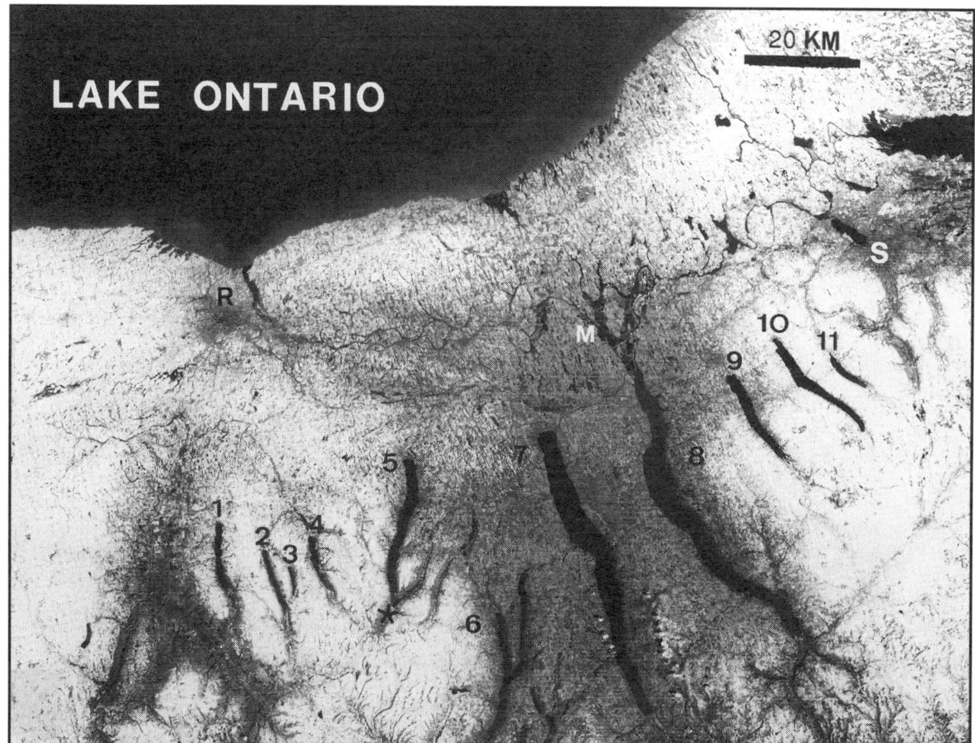

Figure 1. Satellite photograph (EROS Data Center no. E–1234–15244–502) of partially snow-covered Finger Lakes illustrating the 11 lakes: (1) Conesus; (2) Hemlock; (3) Canadice; (4) Honeoye; (5) Canandaigua; (6) Keuka; (7) Seneca; (8) Cayuga; (9) Owasco; (10) Skaneateles; (11) Otisco. S = City of Syracuse; R = City of Rochester; M = Montezuma Wetlands. Large lake to the northeast of Syracuse is Oneida. * = Drillsite.

Canandaigua Valley, thus further restricting the age of the Valley Heads to sometime between ~14.8 and 13.6 ka. In addition, Muller and Calkin (1993) have recently projected downward a series of radiocarbon dates stratigraphically above the Valley Heads to estimate an age of 14.4 ka for the Valley Heads.

South of the Valley Heads is an extensive series of large, outwash-filled meltwater channels (Fig. 2) that formed part of the southward-directed Susquehanna drainage basin. However, today the Finger Lakes drain north to Lake Ontario, with the Valley Heads forming the drainage divides in these "through valleys."

METHODS

Approximately 1,300 km of single channel, high-resolution seismic reflection profiles were collected form the Finger Lakes between 1986 and 1988 using an EG&G Uniboom fired at up to 1,000 J. The Uniboom has a dominant frequency of ~1 kHz, which allows resolution of less than 1 m; when fired at 1,000 J, the Uniboom was able to penetrate the entire Quaternary sediment-fill beneath most of the lakes. Both east-west–oriented transverse profiles with a spacing of 1 km or less, as well as a north-south–oriented axial (longitudinal) profile were collected from each lake. The Uniboom was fired every 1.2 sec, with reflections received by a single eight-element hydrophone streamer, with reflections displayed at sweeps of 0.1 to 0.4 sec. Navigation was by compass heading between known points located on topographic sheets for transverse profiles and by triangulation and cross-ties for longitudinal profiles. Profiles have been correlated with lake shore outcrops (bedrock) in all lakes, as well as with 26 piston cores (up to 5 m long) collected from seismically defined outcrops in Seneca and Cayuga Lakes.

In order to transform time sections measured in the field to depth sections, wide-angle reflection experiments (Bryan, 1980) were conducted in five lakes (Fig. 3) to determine the compressional wave velocity structure of the sediment-fill. Velocities were found to range from 1.5 to 2.1 km/sec, and the six seismically defined depositional sequences beneath the lakes (discussed in the following section) were assigned progressive velocities ranging from 1.5 km/sec for the youngest sequence (VI) to 2.0 km/sec for the oldest sequence (I), with a velocity of 1.455 km/sec used for the water column. Such velocities are in agreement with refraction velocities determined by Faltyn (1957) in Tully Valley, a dry "Finger Lake" valley south of Syracuse. They are also consistent with refraction velocities determined for deposits beneath Alpine glacial lakes where there is a near-linear increase of P-wave velocities with depth ranging from 1.5 to 2.2 km/sec (Finckh et al., 1984). All profiles were digitized, velocity-corrected, and plotted on base maps in order to facilitate the construction of bathymetric,

TABLE 1. FINGER LAKE STATISTICS

Lake	Map Number (Figure 1)	Seismic Profiles (km)	Piston Cores (Number)	Drill Data (Yes/No)	Length (km)*	Maximum Width (km)*	Elevation (m)*	Water Volume ($10^6 m^3$)*	Surface Area (km^2)*
Conesus	1	n.a.	0	No	13	1.3	249	157	14
Hemlock	2	30	0	No	11	0.8	276	106	7
Canadice	3	14	0	No	5	0.6	334	43	3
Honeoye	4	n.a.	0	No	7	1.4	245	35	7
Canandaigua	5	120	0	Yes	25	2.4	210	1,640	42
Keuka	6	180	0	No	32	3.3	218	1,434	47
Seneca	7	400	15	No	57	5.2	136	15,540	175
Cayuga	8	150	11	Yes	61	5.6	116	9,379	172
Cwasco	9	75	0	No	18	2.1	217	781	27
Skaneateles	10	80	0	No	24	3.3	263	1,563	36
Otisco	11	n.a.	0	No	9	1.2	240	78	8

Lake	Drainage Area (km^2)*	DA/SA	Maximum Water Depth (m)	Maximum Sedimentary Thickness (m)	Maximum Erosion Below Lake Level (m)	Maximum Erosion Relative to Sea Level (m)
Conesus	168	12	18*	n.a.	n.a.	n.a.
Hemlock	96	14	29	149	173	+103
Canadice	32	11	27	68	94	+240
Honeoye	95	14	9*	n.a.	n.a.	n.a.
Canandaigua	407	10	84	202	261	-51
Keuka	405	9	57	146	193	+25
Seneca	1,181	7	186	270	442	-306
Cayuga	1,870	11	132	226	358	-242
Owasco	470	17	52	95	140	+77
Skaneateles	154	4	84	140+	255	+8
Otisco	94	12	20*	n.a.	n.a.	n.a.

*Based on Bloomfield, 1978.
n.a. = not applicable
DA/SA = Drainage area to surface area ratio.

isopach, total sediment thickness, and depth to bedrock and sequence contour maps for each lake.

Although seismic reflection profiles were collected from all 11 Finger Lakes, usable data were obtained from only 8. Three relatively small, shallow, eutrophic Finger Lakes (Otisco, Honeoye, Conesus) (Bloomfield, 1978) were found to be largely or completely acoustically impenetrable, which is likely due to the presence of interstitial biogenic methane in near-surface sediments. Interstitial gas greatly reduces the P-wave velocity of the sediment, producing a very strong acoustic impedance contrast at or near the sediment-water interface thus precluding penetration of the acoustic pulse (Schubel and Schiemer, 1973).

Following analysis and mapping of these geophysical data, a 120-m-long drillcore (using wire-line coring techniques) was recovered in 1990 from the dry lake valley 3 km south of Canandaigua Lake (Fig. 1). This drillcore, and the accompanying downhole geophysical data, have provided a long stratigraphic record of the geophysically defined subsurface geology of the Finger Lakes; it is discussed here only on a first-order basis as it relates to the general subsurface stratigraphy of the region. More detailed reports on the Canandaigua drillcore, including downhole geophysics, can be found in other chapters in this volume (Wellner et al., Nobes and Schneider, Wellner and Dwyer).

RESULTS

Our velocity-corrected seismic reflection profiles (Fig. 4) have allowed us to quantify and map the following basic parameters for each lake from which we have usable data: (1) water depth, (2) total sediment thickness, and, (3) depth to bedrock. Mullins and Hinchey (1989) previously reported that the combined transverse and longitudinal profile data reveal similarly shaped basins for the Finger Lakes. At the north end of the lakes, bedrock forms broad, shallow cross sections with relatively thin sediment cover. Water depth and depth to bedrock gradually deepen to the south, reaching maximum values about two-thirds of the way to the south before ascending beneath the southern ends of the lakes, indicating that the Finger Lakes occupy true rock basins. Total sediment thickness also increases to the south where maximum values coincide with maximum depths to bedrock. Maximum water depth (186 m), bedrock erosion

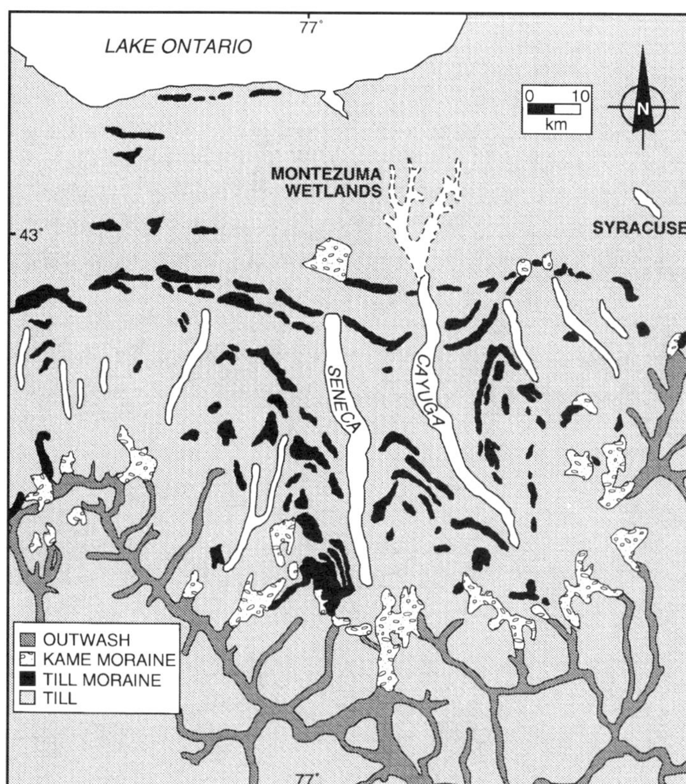

Figure 2. Generalized surficial geologic map of the Finger Lakes region illustrating the distribution of moraines, drumlins, and outwash channels. Note Valley Heads moraines (kame moraines) to the south of the lakes, as well as the chevron nature of till moraines on the interlake uplands. (Based on, and simplified from, Muller and Cadwell, 1986).

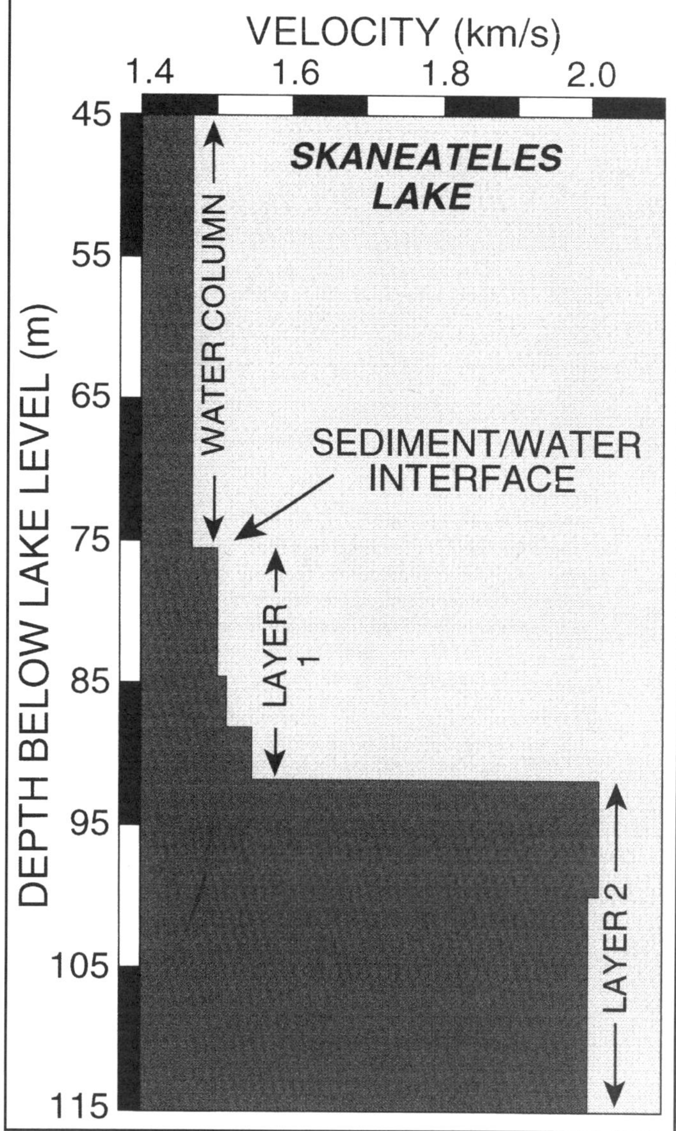

Figure 3. Velocity results of wide-angle reflection experiment conducted in northern Skaneateles Lake using the method of Bryan (1980). Note that velocities of unconsolidated sediments beneath the lake vary from ~1.5 to 2.0 km/sec.

(306 m *below* sea level), and total sediment thickness (270 m) all occur in the southern half of Seneca Lake. However, the Finger Lakes in general are deeply scoured basins with thick sediment-fills (Mullins and Hinchey, 1989).

Our high-resolution seismic reflection profile data have also allowed subdivision of the sediment-fill beneath the Finger Lakes into six first-order depositional sequences based on a combination of reflector terminations and acoustic signature (frequency, amplitude, and continuity). Figure 4 illustrates five of the six sequences (II through VI) on a single profile from central Cayuga Lake. It is these six seismically defined depositional sequences that are the focus of this chapter. The thickness and areal distribution of each sequence has been mapped for each Finger Lake that we have usable data. This has resulted in a very large data base that cannot be presented in its entirety here; rather, our approach is to present and discuss selected results from each lake focusing on the data that are most significant to the regional seismic stratigraphic framework of the Finger Lakes.

Skaneateles Lake

Skaneateles Lake is an intermediate-sized eastern Finger Lake (Fig. 1; Table 1). Based on lead-line sounding, Birge and Juday (1914) reported a maximum water depth of 90.5 m for Skaneateles Lake. However, maximum water depth measured on reflection profiles from Skaneateles Lake is only 84 m (Fig. 5). Our bathymetric map of Skaneateles reveals a steep-sided, flat-floored lake that is deepest in the southern half of the basin (Fig. 5).

Of the six seismically defined depositional sequences identified in the Finger Lakes (I through VI), all are present in Skaneateles Lake with the exception of sequence III. Sequence I (the oldest) is particularly well defined. Figure 6 illustrates the southern third of the axial (longitudinal) profile from Skaneateles, which terminates at the south end of the lake. Sequence I is characterized by chaotic, high-amplitude reflections indicating poorly stratified, coarse-grained deposits (Fig. 6). On transverse profiles, sequence I rests directly on bedrock and its top is highly irregular and hummocky. Isopach mapping of sequence I (Fig. 5) reveals that it is restricted to the southern half of the lake basin and that it thickens to the south where it exceeds 80 m (Fig. 5). This acoustically chaotic, southward thickening wedge of sequence I sediment projects south of the lake to onland outcrops of Valley Heads kame moraine.

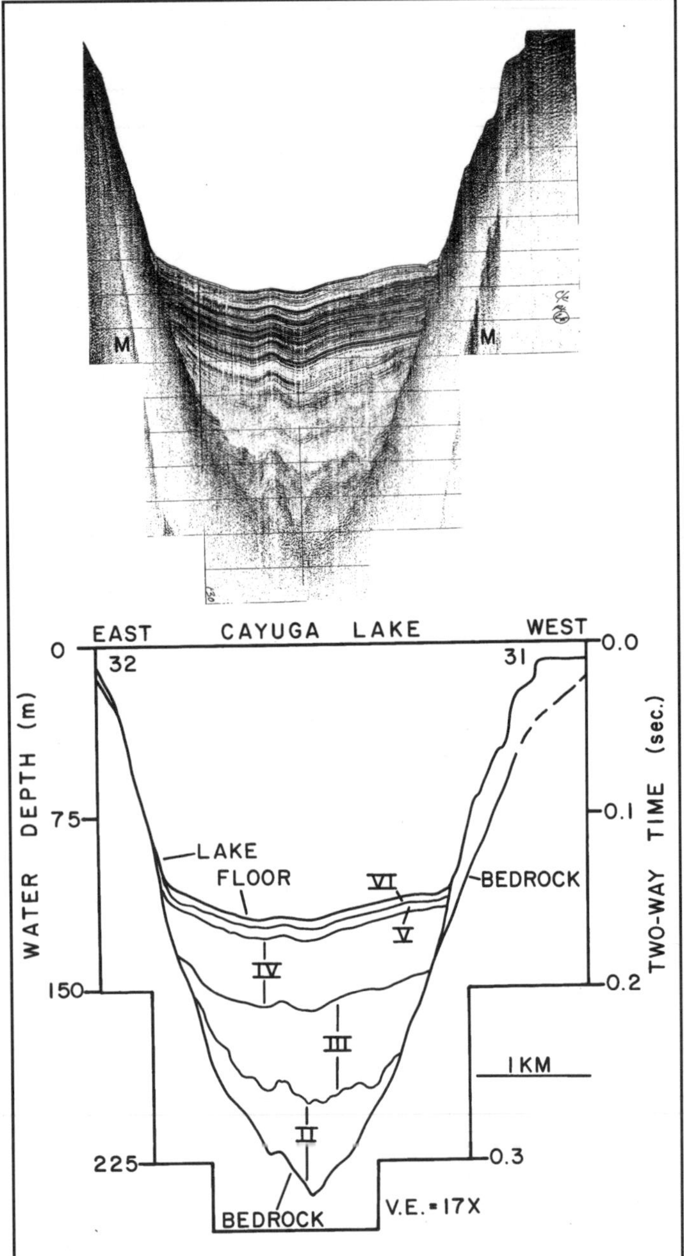

Figure 4. Photograph of a seismic reflection profile and line drawing interpretation from Cayuga Lake illustrating five of the six depositional sequences found beneath the Finger Lakes. Note that these data, once velocity-corrected, allow for measurement of water depth, total sediment thickness, and depth to bedrock. For all profiles displayed in this chapter, M = multiple reflector; V.E. = vertical exaggeration; RX = bedrock reflection. Water Depth assumes a P-wave velocity of 1,500 m/sec (from Mullins and Hinchey, 1989).

Sequence I is overlain and onlapped by sequences II through VI, which are characterized by high-frequency, parallel, continuous reflections suggesting temporally variable but spatially widespread, fine-grained deposits. The fact that these sequences stratigraphically overlay and onlap sequence I indicate that they must be younger than the Valley Heads (i.e., <~14.4 ka). Similar stratigraphic relationships occur in the other Finger Lakes, indicating that the only preserved record of glaciation is that of the last ice advance and retreat. Any pre–late Wisconsin sediments that may have been present beneath the Finger Lakes have been effectively removed by the most recent glacial advance.

Of the more than 140 m of total sediment fill beneath Skaneateles Lake, more than half (80+ m) is represented by depositional sequence II. Sequence II is characterized by strong lateral changes in acoustic facies from chaotic in the north to high-frequency, continuous reflections in the south (Fig. 7). Sequence II also contains four discrete (IIa through IId) wedges (up to 20 m thick) of chaotic to acoustically transparent (reflection-free) deposits that interfinger with high-frequency, continuous reflections. These wedges (IIa through IId) have erosive basal contacts and their tops are highly irregular, yielding large diffraction patterns on transverse profiles (Fig. 8).

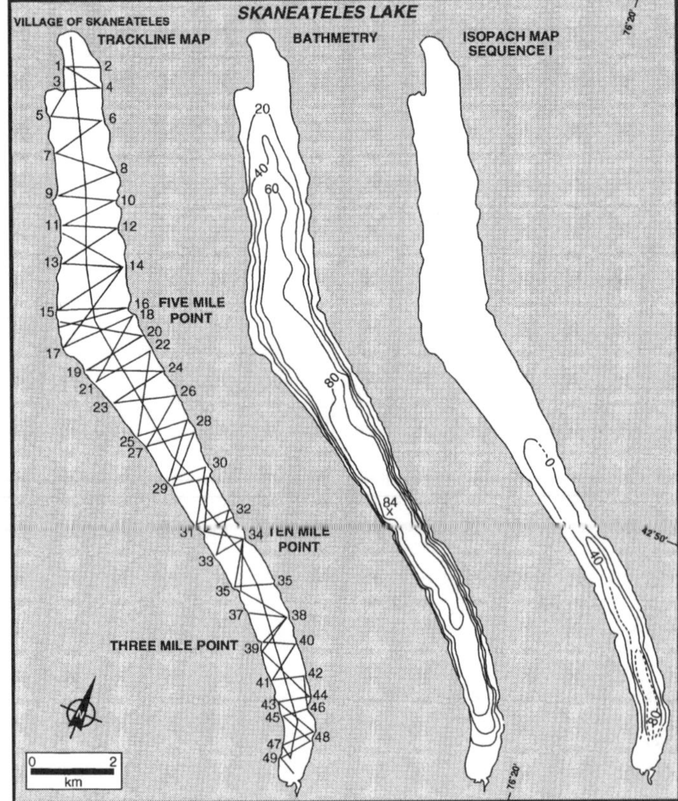

Figure 5. Seismic reflection profile trackline, bathymetric, and sequence I isopach maps for Skaneateles Lake.

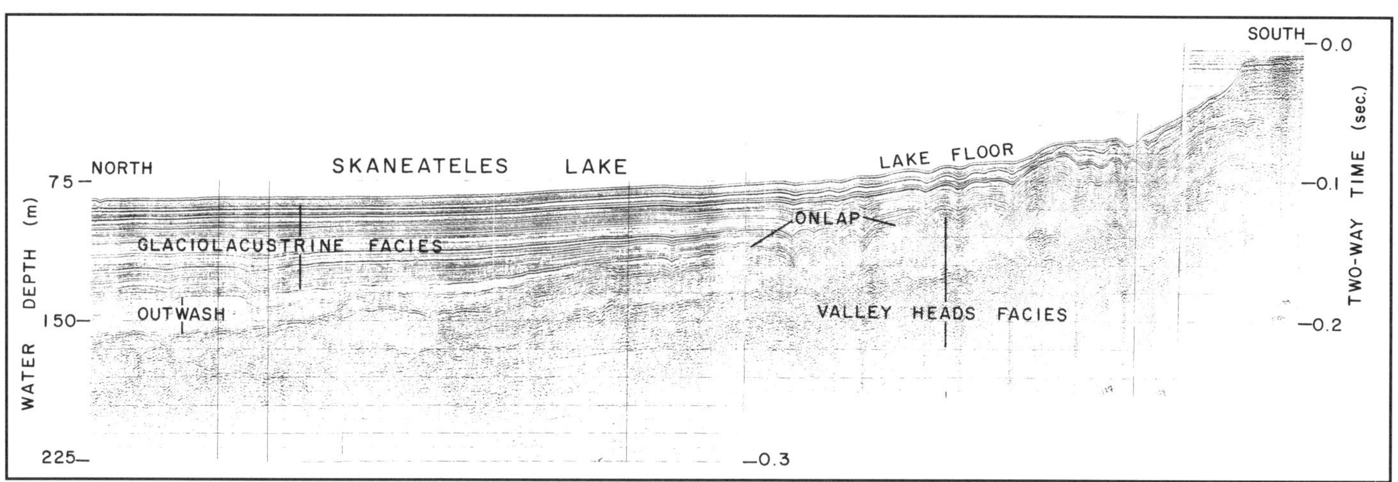

Figure 6. Southern portion (~3 km) of axial seismic reflection profile from Skaneateles Lake. Note acoustically chaotic facies which thickens southward toward outcrops of Valley Heads moraine, and onlap of overlying glaciolacustrine facies which thin toward the south. These stratigraphic relationships indicate that the entire sediment fill beneath Skaneateles Lake is younger than the Valley Heads (i.e., <~14.4 ka).

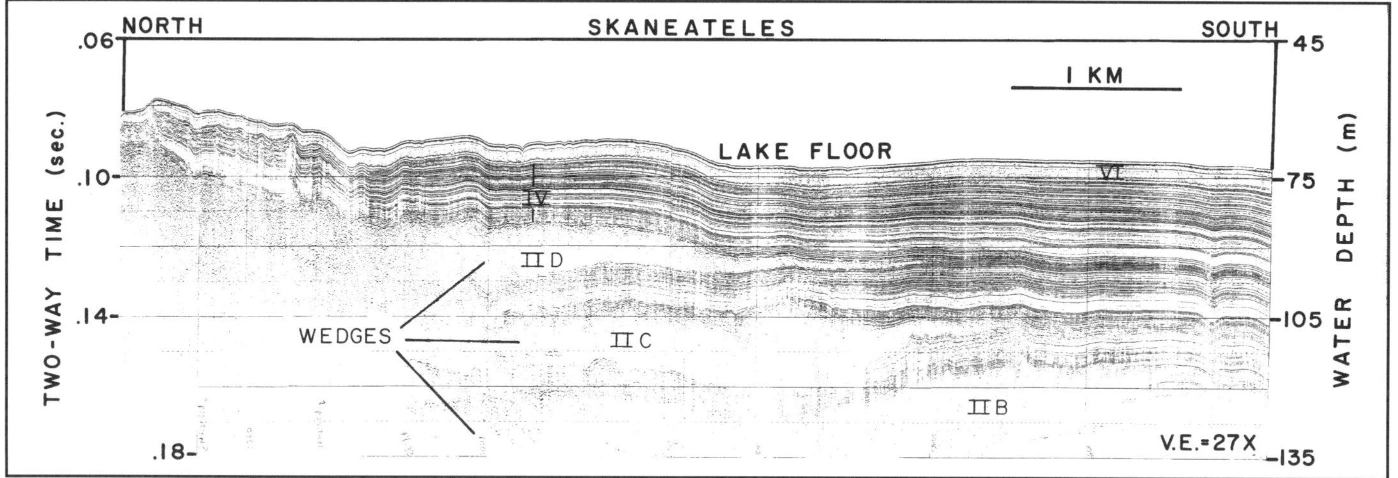

Figure 7. Central portion of axial seismic reflection profile from Skaneateles Lake illustrating transparent (reflection free) wedges of sequence II interbedded with high-frequency, continuous reflections.

The chaotic to transparent wedges in sequence II beneath Skaneateles Lake have not been sampled; thus, their origin is uncertain. However, they are acoustically similar to "till-tongues" described by King et al. (1991) from the Scotian and Norwegian shelves that have been interpreted as grounding line deposits. An alternate interpretation is that these sequence II wedges represent the abrupt, massive outpourings of subglacial sediment into a proglacial lake. Whether these wedges are till-tongues or sediment outpourings cannot be resolved by our present data base. However, both interpretations would require rapid retreat of the ice margin from its Valley Heads position south of Skaneateles Lake to about Five Mile Point where it became pinned. If these wedges are till-tongues, it would imply a number of small-scale oscillatory advances and retreats of the ice margin to produce the observed acoustic stratigraphy. Alternatively, if they are massive outpourings of sediment, it would imply the occasional sudden release of large volumes of subglacial sediment.

Beneath Skaneateles Lake, sequence II is overlain directly by high-frequency, continuous reflections of sequence IV (Figs. 7 and 8). Individual reflections within sequence IV can be traced along nearly the entire length of Skaneateles Lake, implying a second phase of rapid ice margin retreat from its sequence II position to the northern end of the lake.

Owasco Lake

Owasco Lake is a relatively small eastern Finger Lake, just west of Skaneateles Lake (Fig. 1). South of Owasco is an extensive dry lake valley with outcrops of Valley Heads kame moraine occurring ~16 km south of the lake (Fig. 2). Because of

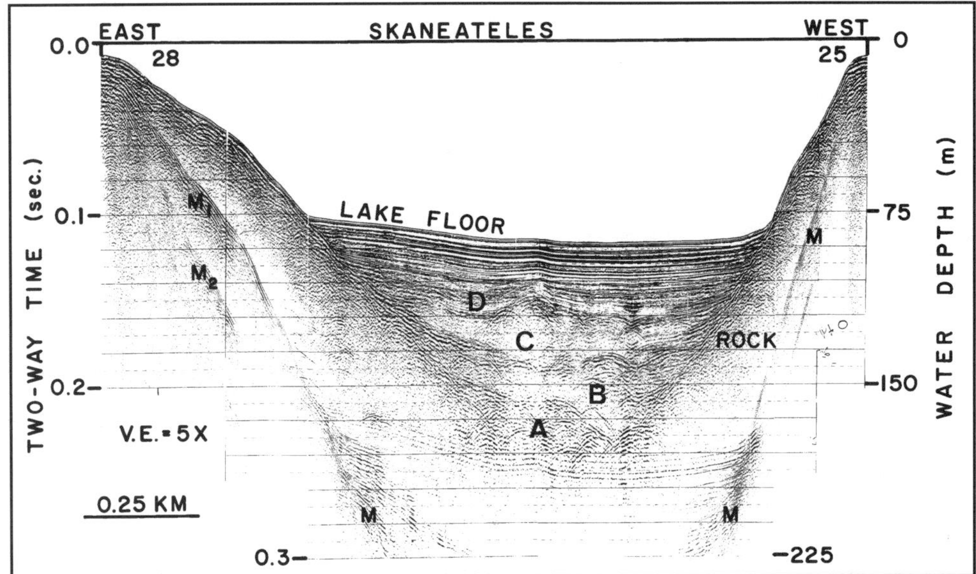

Figure 8. Transverse seismic reflection profile 28-25 (see Fig. 5) illustrating buried wedges (A–D) shown in Figure 7. Note large diffraction patterns at tops of wedges indicating highly irregular surfaces.

these relationships, Owasco Lake has the largest drainage area to surface area ratio (17) of any Finger Lake (Table 1). As a consequence, the youngest depositional sequences (V and VI) in Owasco Lake are particularly well developed.

Our bathymetric map of Owasco (Fig. 9) reveals a rather symmetrical north-south lake basin, which is steep-sided and flat-floored. Maximum water depth (52 m) occurs at about the mid-point of the lake basin with relatively gradual ascents at both the north and south ends of the lake (Fig. 9). Of the six depositional sequences identified in the Finger Lakes, five (II through VI) have been identified beneath Owasco Lake. Sequence I (Valley Heads equivalent) was not found beneath Owasco Lake most likely because of the extensive dry lake valley south of the lake and distance (~16 km) from known Valley Heads outcrops.

One of the more intriguing features beneath Owasco Lake is a buried ridge at the north end of the lake, which is part of depositional sequence II (Fig. 10). This ridge has as much as 25 m of relief and is up to 0.5 km wide at its base. The feature occurs between profiles 3-4 and 14-17 (a distance of about 2.5 km) as a sinuous ridge along the longitudinal axis of the lake. Acoustically, the ridge is characterized by chaotic to discontinuous high-amplitude reflections, which suggests that is consists of poorly stratified, coarse-grained sediment that appears to rest directly on bedrock. It is our interpretation that this ridge is a buried esker, which implies the presence of a large subglacial meltwater tunnel beneath the ice sheet during, and perhaps prior to, deposition of sequence II in Owasco Lake.

Approximately half the maximum total sediment thickness (95 m) beneath Owasco Lake consists of depositional sequence IV. This sequence extends the entire length of the lake basin and

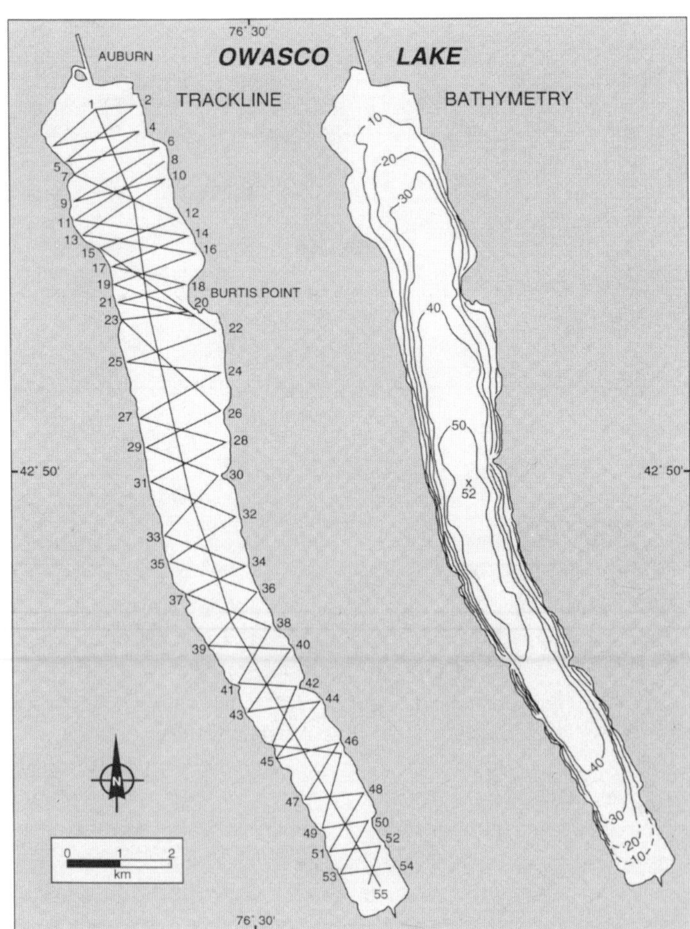

Figure 9. Seismic reflection profile trackline and bathymetric maps of Owasco Lake.

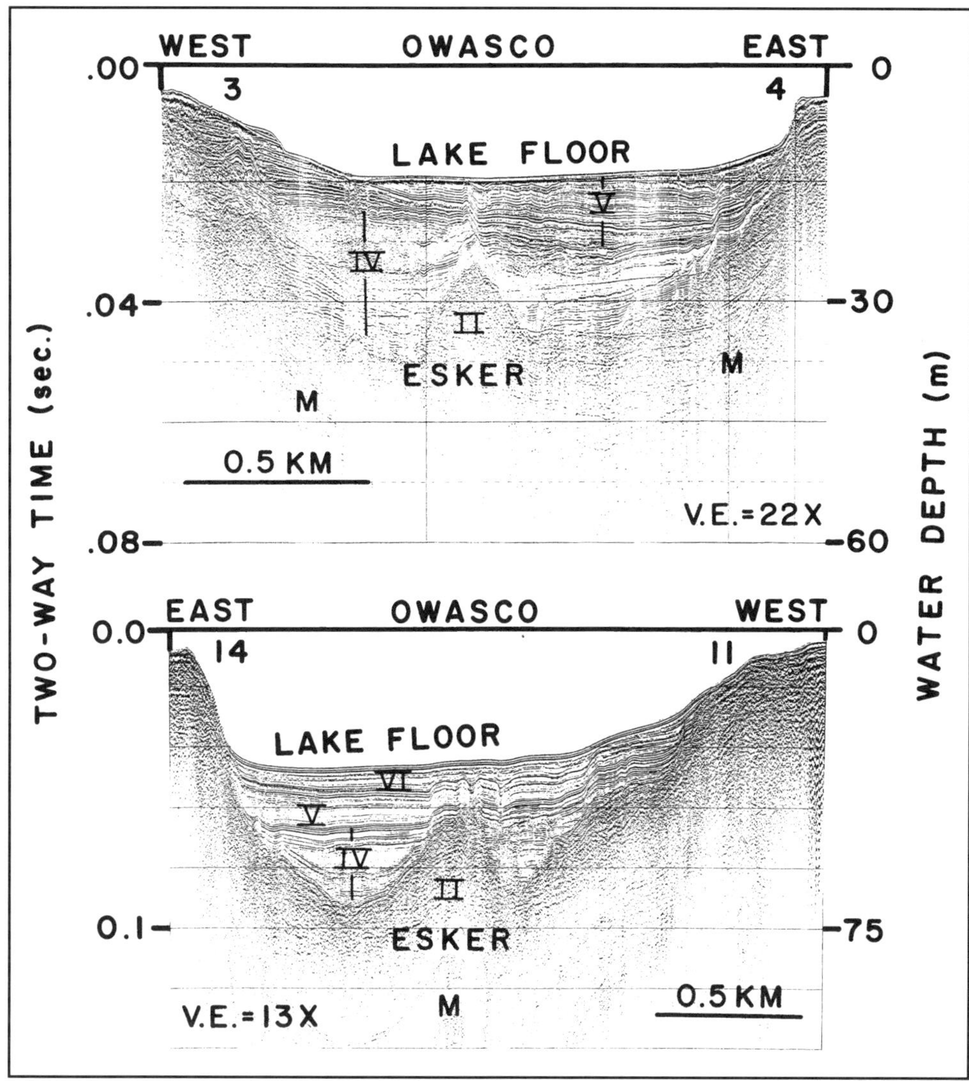

Figure 10. Transverse seismic reflection profiles 3-4 and 14-11 (see Fig. 9) from Owasco Lake illustrating buried ridge interpreted as an esker.

is characterized by high-frequency continuous reflections (Fig. 11). These highly continuous reflections of sequence IV are interrupted in the northern half of Owasco Lake by an acoustically transparent (reflection-free) wedge of sediment (Fig. 11) up to 15 m thick and extending longitudinally (north-south) for a distance of 7 km (Fig. 12). The lower contact of this wedge shows evidence of minor erosion, and its top is flat and conformable (Fig. 11). The lack of internal impedance contrasts (i.e., transparency) further suggests that the wedge consists of massive (i.e., homogenous) fine-grained sediment. The Owasco wedge is unlike the chaotic to transparent wedges of sequence II beneath Skaneateles Lake in that it lacks an irregular surface, which gives rise to large diffraction patterns in Skaneateles (Fig. 8). It is our interpretation that the transparent wedge within sequence IV beneath Owasco Lake represents the sudden discharge of fine-grained sediment (inflow) from beneath the Laurentide ice sheet into a high-standing proglacial lake.

Stratigraphic relationships and sequence isopach maps from the Finger Lakes in general indicate that depositional sequences I through IV were derived from a northerly source (i.e., the Laurentide ice sheet) with overflow drainage to the south. Sequence V, however, displays clear evidence for a southerly (as well as lateral) source of sediment. Beneath Owasco Lake, sequence V is thickest (21 m) at the south end of the lake (Fig. 12) where the modern Owasco Inlet flows into the lake. Sequence V also thickens abruptly toward Burtis Point (Fig. 13), which is the largest delta along the Owasco lakeshore. Muller and Cadwell (1986) mapped lacustrine deposits along the col connecting Burtis Point and Skaneateles Lake that suggests the possible overflow of Skaneateles proglacial lake waters into Owasco Lake to build Burtis Point during sequence V. This seismic stratigraphic evidence for southerly and laterally derived sediment input to the Finger Lakes argues for an abrupt drainage reversal at the beginning of sequence V. Such a

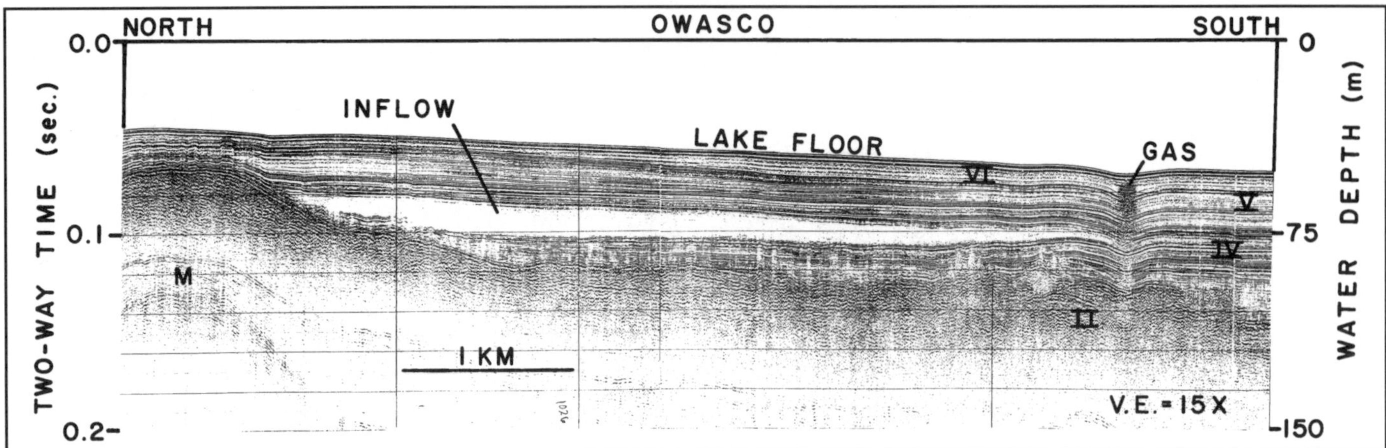

Figure 11. North-central portion of axial seismic reflection profile from Owasco Lake illustrating transparent (reflection free) wedge of sediment within sequence IV interpreted as a sudden "inflow" of sediment.

reversal probably occurred in response to the lowering of high-standing proglacial lake levels in the Finger Lakes region (Fairchild, 1909; Muller and Prest, 1985) due to northward ice retreat and the initiation of modern north-directed drainage of the Finger Lakes.

Because of the large drainage area to surface area ratio for Owasco Lake (Table 1), the youngest depositional sequence (VI) is very well developed. Sequence VI consists of low-amplitude, discontinuous internal reflections, suggesting fine-grained sediments derived from multiple point sources. These include modern, postglacial sediments found throughout the lake basin, except along the steep lateral slopes of the lake where sequence VI is very thin or absent. Sequence VI reaches a maximum thickness of 11 m (Fig. 12) near the midpoint of the Owasco Lake basin coincident with maximum water depth (cf. Figs. 9 and 12). This relationship strongly suggests that postglacial sediments are being "focused" to the deepest part of the lake basin via resuspension and redeposition, which is typical of many modern lakes (Lehman, 1975; Davis and Ford, 1982).

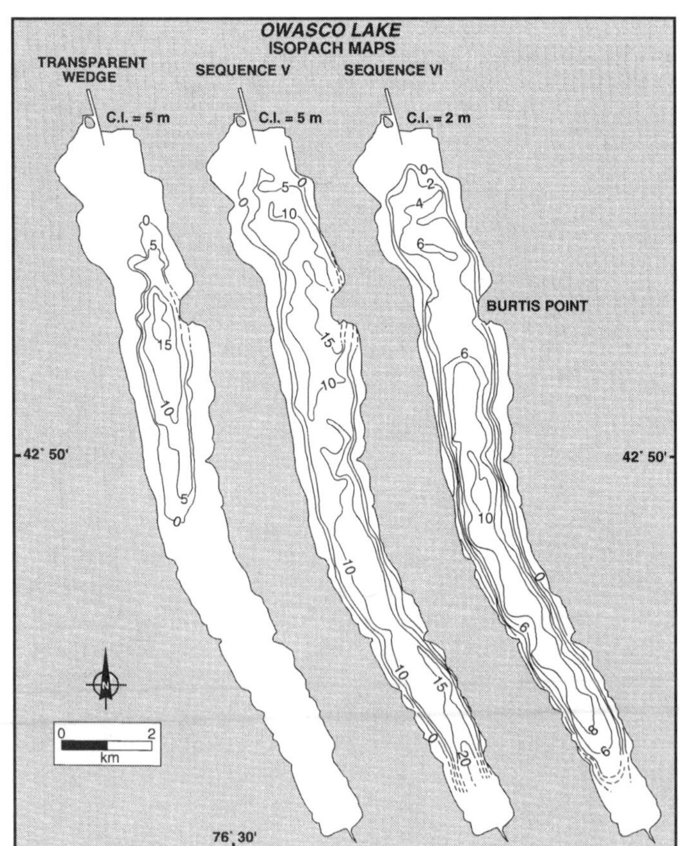

Figure 12. Isopach maps of the transparent wedge illustrated in Figure 11; sequence V; and sequence VI in Owasco Lake. Note that transparent wedge is restricted to the northern portion of the lake basin; that sequence V thickens to both the south and toward Burtis Point, and that sequence VI sediments are thickest in the deepest portion of the lake. C.I. = contour interval.

Cayuga Lake

Cayuga Lake is one of the two largest Finger Lakes (Fig. 1). Fairchild (1934a) referred to Cayuga Lake as the "axial depression of central New York" because it has the lowest lake level elevation (Table 1) in the region. It has also long been known that the lake floor of Cayuga extends below sea level (Tarr, 1894).

The valley walls and uplands surrounding Cayuga Lake contain numerous hanging valleys and deltas (Tarr, 1904; Bloom, 1986), attesting to large-scale glacial overdeepening of the valley and high standing proglacial lakes. Along the southwest margin of the lake at "Fernbank" there is evidence for interglacial deposits of Sangamon(?) age (Maury, 1908; von Engeln, 1929; Karrow et al., 1990), which has been cited as evi-

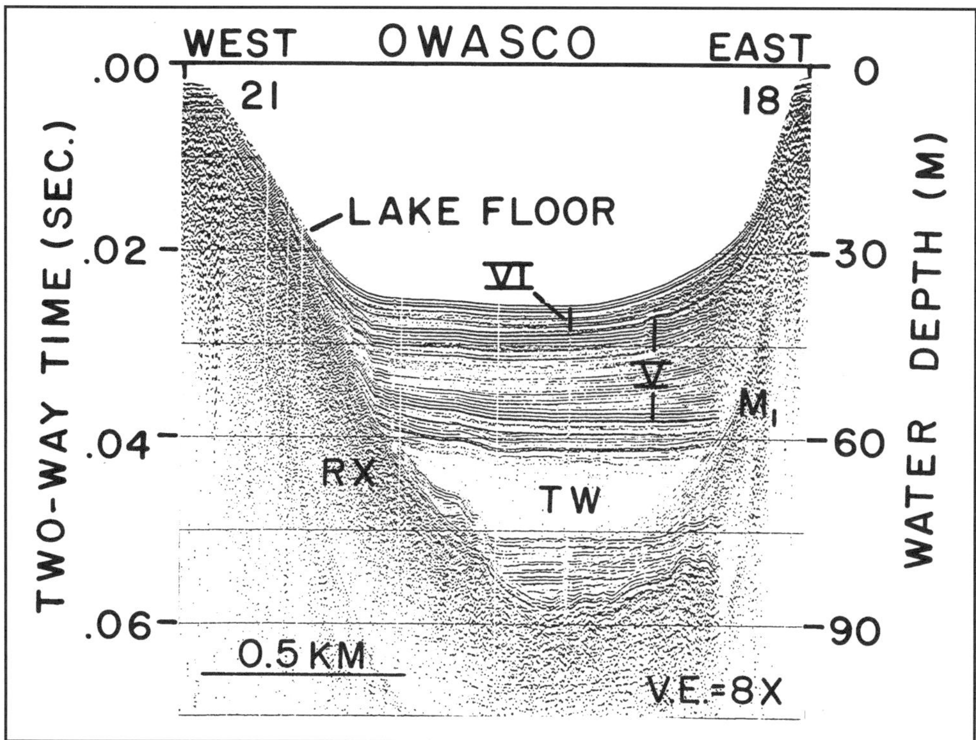

Figure 13. Transverse seismic reflection profile 21-18 (see Fig. 9) from Owasco Lake illustrating the transparent wedge in sequence IV, as well as the thickening of sequence V to the east toward Burtis Point. RX = bedrock reflection, TW = transparent wedge, M = multiple reflection.

dence for multiple glaciations and excavations of Cayuga Valley (Bloom, 1984, 1986).

Approximately 150 km of seismic reflection profiles have been collected from Cayuga Lake along with 11 piston cores (Fig. 14). Our bathymetric map of Cayuga reveals that the northern end of the lake is relatively shallow (<20 m) but gradually deepens to the south, where a maximum water depth of 132 m occurs in the southern half of the lake basin, before shoaling at the south end of the lake (Fig. 14). In transverse profile, Cayuga is a typical Finger Lake, being steep-sided and flat-floored (Fig. 4), particularly in the southern two-thirds of the lake (Fig. 14).

The bedrock profile of Cayuga Lake is similar to its bathymetric profile, being relatively shallow in the north and deepening southward before rising beneath the south end of the lake (Fig. 15). Maximum depth of erosion below lake level (358 m) occurs in the southern half of the lake basin (Fig. 15), where it extends as much as 242 m *below* sea level. Reflection profiles from the southern half of the lake (Fig. 16) indicate that the bedrock morphology is V-shaped. Drilling at the south end of the lake basin in the city of Ithaca revealed that bedrock extends only 133 m below lake level (17 m *below* sea level; Coates, 1968), indicating that Cayuga Lake occupies a true rock basin (Tarr, 1894).

Glacial erosion beneath Cayuga Lake extended through relatively soft Devonian Hamilton shales and siltstones and into the more resistant underlying Onondaga Limestone (Fig. 17). Erosion appears to have approximately followed the down-dip contact between Hamilton shales and Onondaga limestones before rising up through shales at the south end of the lake basin. Total sediment fill follows the overall longitudinal bedrock profile being relatively thin in the north and thickening to a maximum of 226 m in the southern half of the lake basin before thinning beneath the south end of the lake (Fig. 15). Our isopach map of the total sediment fill beneath Cayuga Lake also defines a series of closed basins along the longitudinal axis of Cayuga Lake (Fig. 15).

All six Finger Lake depositional sequences are present beneath Cayuga Lake (Figs. 4 and 16). As in Skaneateles Lake, sequence I is restricted to the southern half of the basin where it consists of high-amplitude chaotic reflections which thicken to the south where it exceeds 100 m. Sequence II is a thick unit (up to 150 m) exhibiting strong lateral changes in acoustic facies from hummocky, high-amplitude reflections in the north (Fig. 4) to semi-continuous, low-amplitude reflections (Fig. 16) in the south where it onlaps sequence I. Sequence III is a relatively thin (<50 m) unit characterized by a reflection-free to low-amplitude, discontinuous seismic facies.

Depositional sequence IV is particularly well-developed in Cayuga Lake. This sequence is characterized by a high-frequency, continuous acoustic facies (Figs. 4 and 16) with individual reflections extending nearly the entire length (60 km) of

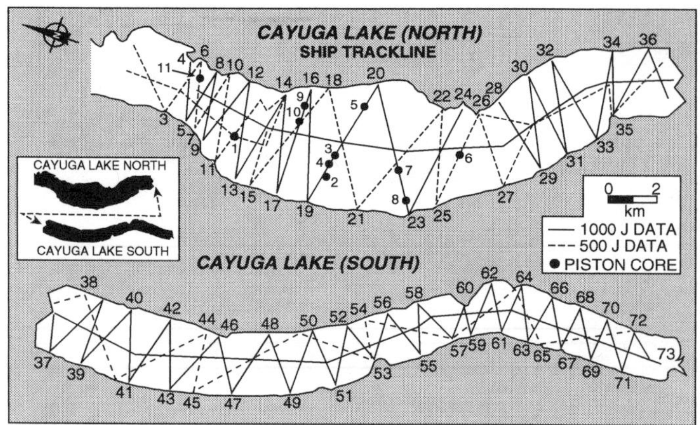

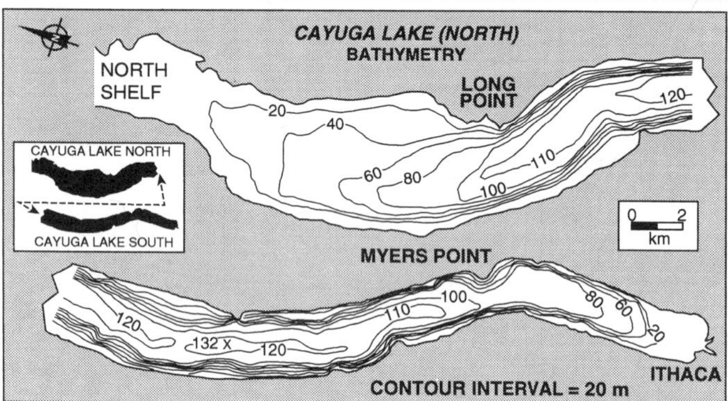

Figure 14. Seismic reflection profile trackline and bathymetric maps of Cayuga Lake. Note that for purposes of presentation, all maps of Cayuga Lake have been divided into northern and southern halves, with north oriented to the left. Locations of piston cores also shown on trackline map.

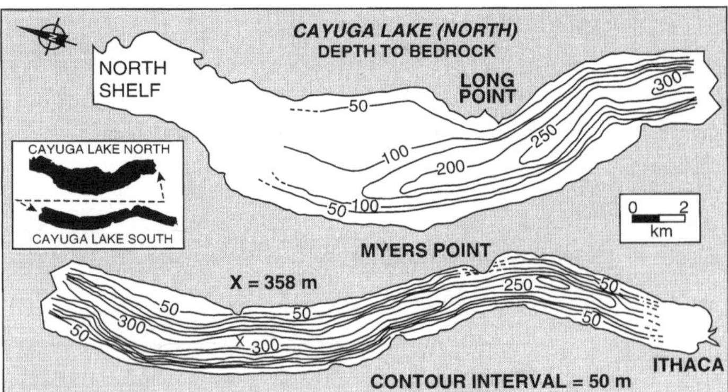

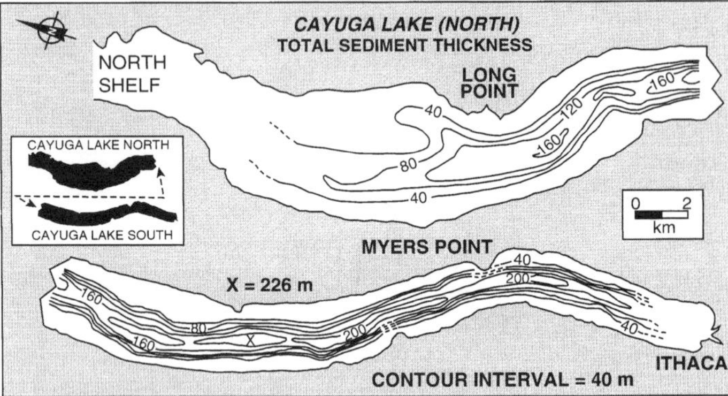

Figure 15. Structure contour map of the eroded bedrock surface and total sediment thickness map for Cayuga Lake.

the lake basin. Sequence IV has a maximum measured thickness of 66 m at the north end of the lake where it forms a northward thickening wedge of sediment (Fig. 18). This wedge probably represents the outpouring of sediment from the adjacent Montezuma wetlands (Fig. 2), which appear to be a system of subglacial meltwater channels (see Petruccione et al., this volume). The flushing of sediment from Montezuma channels would also explain why the north end of Cayuga Lake is relatively shallow. At the south end of Cayuga Lake, sequence IV thins and begins to onlap underlying sequences I and II. However, throughout much of the length of Cayuga Lake, sequence IV maintains a relatively uniform thickness on the order of ~30 m. The fact that sequence IV reflections can be traced for tens of kilometers along nearly the entire length of Cayuga Lake, coupled with its abrupt thickening at the north end, indicates its sediment source lay to the north and suggests that the ice margin was located along till moraines (Fig. 2) near the north end of Cayuga Lake during sequence IV deposition.

Unlike sequence IV, which was largely derived from the north, sequence V displays distinct evidence of being derived from the *south* end of Cayuga Lake (Fig. 18). Sequence V exceeds 35 m in thickness at the south end of the lake near Ithaca, as well as off Meyers Point where it thickens abruptly (Fig. 18). The fact that underlying sequences do not display evidence of thickening indicates that Meyers Point was built into Cayuga Lake during sequence V. This abrupt change in sediment source argues for a drainage reversal at the beginning of sequence V. As in Owasco Lake, this drainage reversal most likely was a response to a drop in lake level, once ice had withdrawn from the north end of Cayuga Valley opening lower elevation outlets to the east and establishing the "modern" north-directed drainage of the Finger Lakes.

Sequence VI is the youngest seismic unit in Cayuga Lake. It is characterized by a low-amplitude, discontinuous to reflection-free acoustic facies. This sequence is thickest (~12 m) in the southern half of the lake basin, as well as off major tributary deltas ("points"), indicating a southerly source of sediment, lateral stream input, and the focusing of modern sediment to the deep lake floor.

Depositional sequences II, IV, and VI have been sampled beneath Cayuga Lake by piston coring. Eleven cores were recovered from seismically defined outcrops in the northern half of the lake basin (Figs. 14 and 19). An esker-like ridge of sequence II was sampled by core 10 (Fig. 19), which recovered rounded gravels overlain by a thin (<1 m) deposit of fine-grained rhythmites with dropstones. This core confirms that at the north end of Cayuga Lake sequence II consists of coarse-grained sediment.

Sediments of sequence IV consist of very fine-grained,

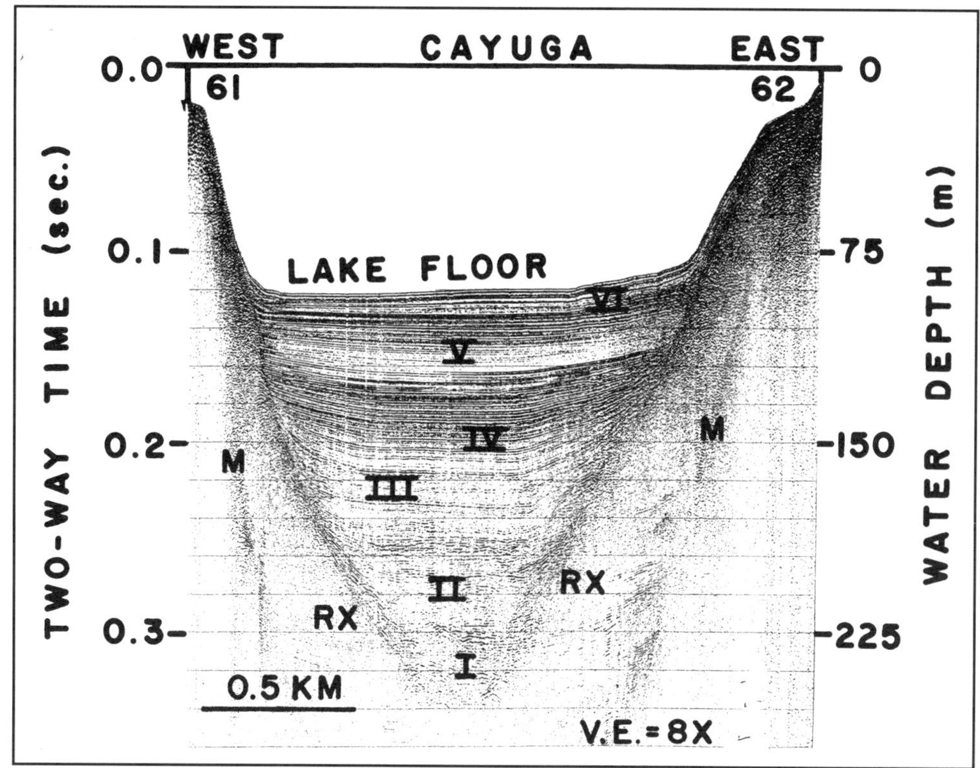

Figure 16. Transverse seismic reflection profile 61-62 (see Fig. 14) from southern Cayuga Lake illustrating eroded bedrock surface and depositional sequences that constitute the sediment fill.

organic-poor, pink-clay rhythmites with occasional dropstones. X-radiographs reveal distinct laminations that vary in thickness from 1 to 10 cm. We interpret these rhythmic (varve?) sediments as proglacial lake deposits that give rise to a high-frequency, continuous acoustic facies that likely is a composite of many small-scale impedance contrasts.

In the profundal zone (deep lake floor), sequence VI sediments consist of dark gray to black, relatively organic-rich muds. X-radiographs indicate that these sediments are either weakly or well-laminated. In contrast, sequence VI littoral zone (nearshore) sediments are coarser grained massive sands and carbonate-rich marls.

Ludlam (1967) examined 72 piston cores (average length, 3 m) from the southern half of Cayuga Lake at water depths exceeding 45 m. All of his cores were recovered from our sequence VI, which significantly exceeds 3 m in thickness throughout the southern half of Cayuga Lake. Ludlam found that the near-surface sediments are distinctly "banded," with centimeter-scale couplets consisting of dark, fine-grained bands separated by light silt and clay layers. Ludlam (1967) interpreted these couplets as annual layers representing spring flood and nonflood conditions.

Additional insight into the lithologic nature of depositonal sequences beneath Cayuga Lake comes from 13 drill records reported by Tarr (1904) from the dry lake valley south of Cayuga Lake in the city of Ithaca. These wells were drilled to subsurface depths of 71 to 107 m, with only two of the wells

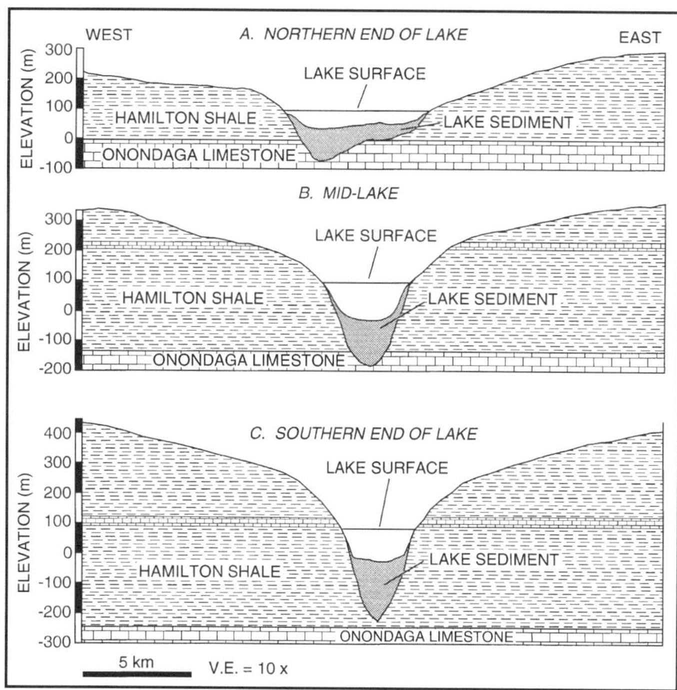

Figure 17. Schematic transverse profiles of the (A) northern, (B) central, and (C) southern portions of Cayuga Lake illustrating water depths, sediment-fill, and eroded bedrock surface superimposed on subsurface Devonian bedrock.

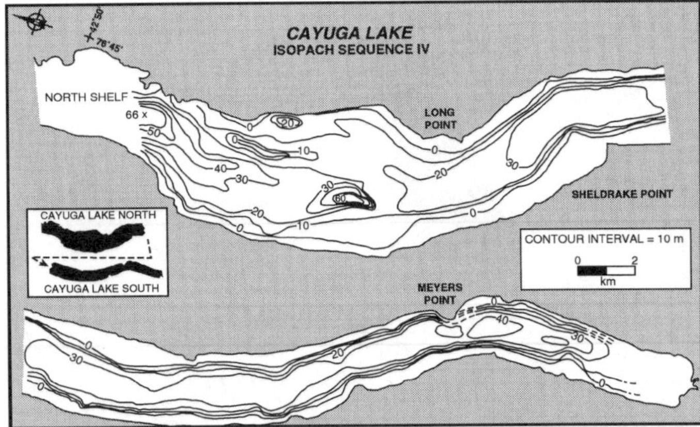

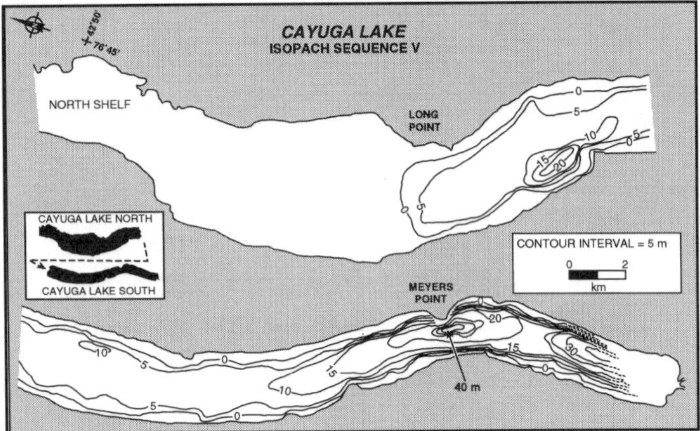

Figure 18. Isopach maps of sequences IV and V in Cayuga Lake. Note thickest accumulations at north end of lake for sequence IV, and southern end of lake for V.

penetrating bedrock. The stratigraphy in these 13 wells is similar, revealing four first-order units. The basal deposit is a heterogeneous, coarse-grained unit from which artesian water flowed at rates of up to 300,000 gallons/day. Tarr (1904) described this basal unit as consisting of variable sediment ranging from washed sand and gravel to "till," to "quicksand." These coarse basal deposits are overlain by "a great thickness of clay" (35 to 65 m thick), which is devoid of fossils but does contain occasional "scratched, angular pebbles" (Tarr, 1904). Overlying this clay unit is a coarse-grained deposit 6 to 21 m thick. This upper coarse unit consists of "well-washed" sands and "well-rounded" gravels with occasional plant fragments, mollusks, and logs. The uppermost unit in these 13 wells is predominantly a fine-grained massive clay up to 18 m thick with abundant fragments of mollusks, plants, and wood, including logs.

It was Tarr's (1904, p. 76) view that all four of these subsurface units are of late Wisconsin age: "Neither here nor in the other well that reached rock, nor, in fact in any of the wells, was any older drift encountered. All the materials are such as might have been brought by the last ice advance, or deposited since the ice-sheet melted away." Tarr (1904) also interpreted the basal coarse unit as a subsurface extension of the nearby Valley Heads Moraine that we correlate with depositional sequence I beneath Cayuga Lake. A land-based multichannel seismic reflection profile shot across Cayuga Valley at Ithaca (Mullins et al., 1991b) also revealed four first-order stratigraphic units, with the basal (Valley Heads) unit up to 80 m thick.

The middle clay layer in these 13 wells was interpreted by Tarr (1904) as a proglacial lacustrine sequence. The "scratched angular stones" he referred to may be dropstones, which would imply the presence of icebergs and a calving ice margin fronting in a large proglacial lake. We correlate this clay unit with sequence IV beneath Cayuga Lake, which is consistent with results from our piston cores.

The overlying washed sands and gravels drilled at Ithaca may be a fluvial/alluvial or shallow lacustrine facies that would record a drop in lake level, drainage reversal, and the outpouring of coarse debris from nearby glens and gorges. We interpret this upper coarse unit as a lateral facies equivalent to southward-thickening sequence V beneath Cayuga Lake. The succeeding organic-rich clays at the top of the wells are likely a postglacial, shallow lacustrine facies that would be equivalent to our sequence VI beneath the lake. Tarr (1904) suggested that the now-dry lake valley south of Cayuga Lake was flooded due to differential isostatic rebound and tilting following retreat of the late Wisconsin ice sheet.

Seneca Lake

Seneca is the largest of the 11 Finger Lakes (Table 1). Lead-line soundings of Seneca Lake made in the late 1800s revealed that the bottom of the lake extends as much as 52 m *below* sea level (Schaffner and Oglesby, 1978), and, by the early 1900s, borehole results had documented bedrock erosion exceeding 200 m *below* sea level south of the lake near Watkins Glen (Coates, 1968). This led Fairchild (1934b), who was an ardent opponent of the glacial theory for the origin of the Finger Lakes, to refer to Seneca Valley as the "Susqueseneca Grand Canyon" of the east.

About 20 yr prior to our initial seismic reflection study of Seneca Lake (Fig. 20) (Stephens, 1986; Mullins and Hinchey, 1989), the U.S. Navy contracted Lamont-Doherty Geological (Earth) Observatory of Columbia University to conduct a geophysical site survey of central Seneca Lake. Woodrow et al. (1969) published a line drawing of one of these reflection profiles confirming early borehole evidence of deep bedrock scour. Those authors (Woodrow et al., 1969) also argued that there had been "a complex series of lake-filling events" in the Seneca Lake basin.

Our bathymetric map of Seneca Lake (Fig. 20) reveals a longitudinally asymmetrical basin that gradually deepens from north to south before shoaling at the south end of the lake. Maximum water depth of 186 m occurs in one of two closed basins in the southern half of the lake (Fig. 20). In transverse profile, Seneca Lake is relatively steep-walled and flat-floored (Figs. 20 and 21). Onlap of the sediment-fill against the steep

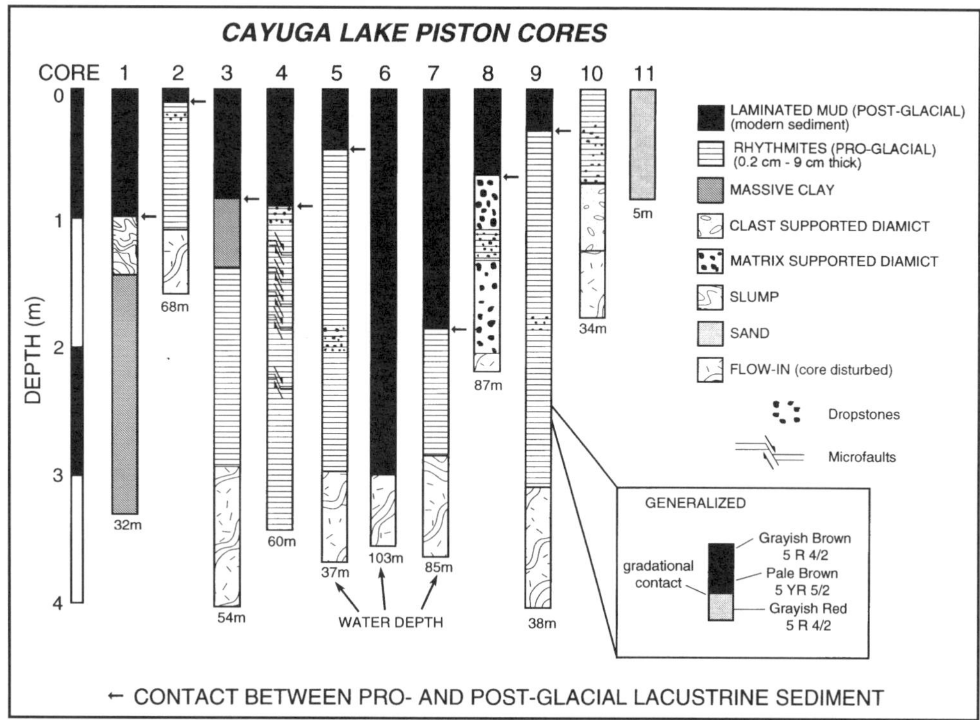

Figure 19. Schematic, lithologic descriptions of piston cores recovered from Cayuga Lake (see Fig. 14 for locations).

bedrock sides of Seneca Lake produces a distinct U-shaped profile. However, tracing of the bedrock reflection beneath the sediment fill reveals a more V-shaped profile of the eroded bedrock surface (Fig. 21).

The eroded bedrock surface beneath Seneca Lake (Fig. 22) broadly mimics the distribution of water depth (Fig. 20). Bedrock deepens from north to south, reaching a maximum of 442 m below lake level (306 m *below* sea level) in the southern half of the lake basin (Fig. 22). There is also an overdeepened erosional trough along the axis of Seneca Lake, as well as a series of closed bedrock depressions (Fig. 22). Although very hummocky bottom topography at the south end of Seneca Lake did not allow us to map the bedrock surface here, onland data document a southward rise of the bedrock surface. A multichannel weight-drop reflection profile collected from the eastern half of the dry lake valley 3.5 km south of Seneca Lake (Fig. 23) indicates that bedrock extends at least 240 m below lake level; drillhole data from near the south shore of Seneca Lake indicate bedrock erosion to depths of 281 m below lake level (W. McPherson, U.S. Geological Survey, Ithaca, personal communication, 1991). Thus, bedrock at the south end of Seneca Lake extends only about 145 m below sea level compared to a maximum of 306 m below sea level beneath the lake itself, indicating that the eroded bedrock surface does rise to the south and that Seneca Lake occupies a rock basin.

In addition to being the largest and most deeply scoured Finger Lake, Seneca Lake also contains the thickest sediment-fill. A maximum of 270 m of sediment occurs in the southern half of the lake basin (Fig. 22). Similar to water depth and depth to bedrock distributions (Figs. 20 and 22), total sediment thickness along the axis of the lake is at a minimum in the north end of the basin, thickens southward, and then thins toward the south end of the lake. Thickest sediment accumulations occur in a series of closed, axial basins that coincide with isolated bedrock depressions (Fig. 22). As with the other Finger Lakes, total sediment thickness is very thin to absent along the steep, water-covered bedrock walls of the lake basin (Fig. 21).

All six seismically defined Finger Lake depositional sequences occur in Seneca Lake (Fig. 21). The distributions of each of these six sequences is similar to that found in the other Finger Lakes. However, sequences II and IV contain special features that are deserving of expanded discussion.

Sequence II occurs along nearly the entire length of Seneca Lake, with a maximum thickness of 95 m in the northern half of the lake basin (Fig. 24). However, sequence II is restricted to the overdeepened, axial trough of Seneca Lake (Figs. 24 and 25). As in the other Finger Lakes, sequence II is characterized by strong lateral changes in acoustic facies from chaotic, high-amplitude reflections in the north to semi-continuous, low-amplitude reflections in the south. In addition, two buried ridges are found within sequence II in the northern half of Seneca Lake. These ridges are up to 6 km long, as much as 0.5 km across, and have up to 45 m of relief (Figs. 24 and 25). The sediments that make up these ridges have a chaotic acoustic facies and rest directly on bedrock in the deepest thalweg of the overdeepened, axial trough of Seneca Lake. Similar to Owasco

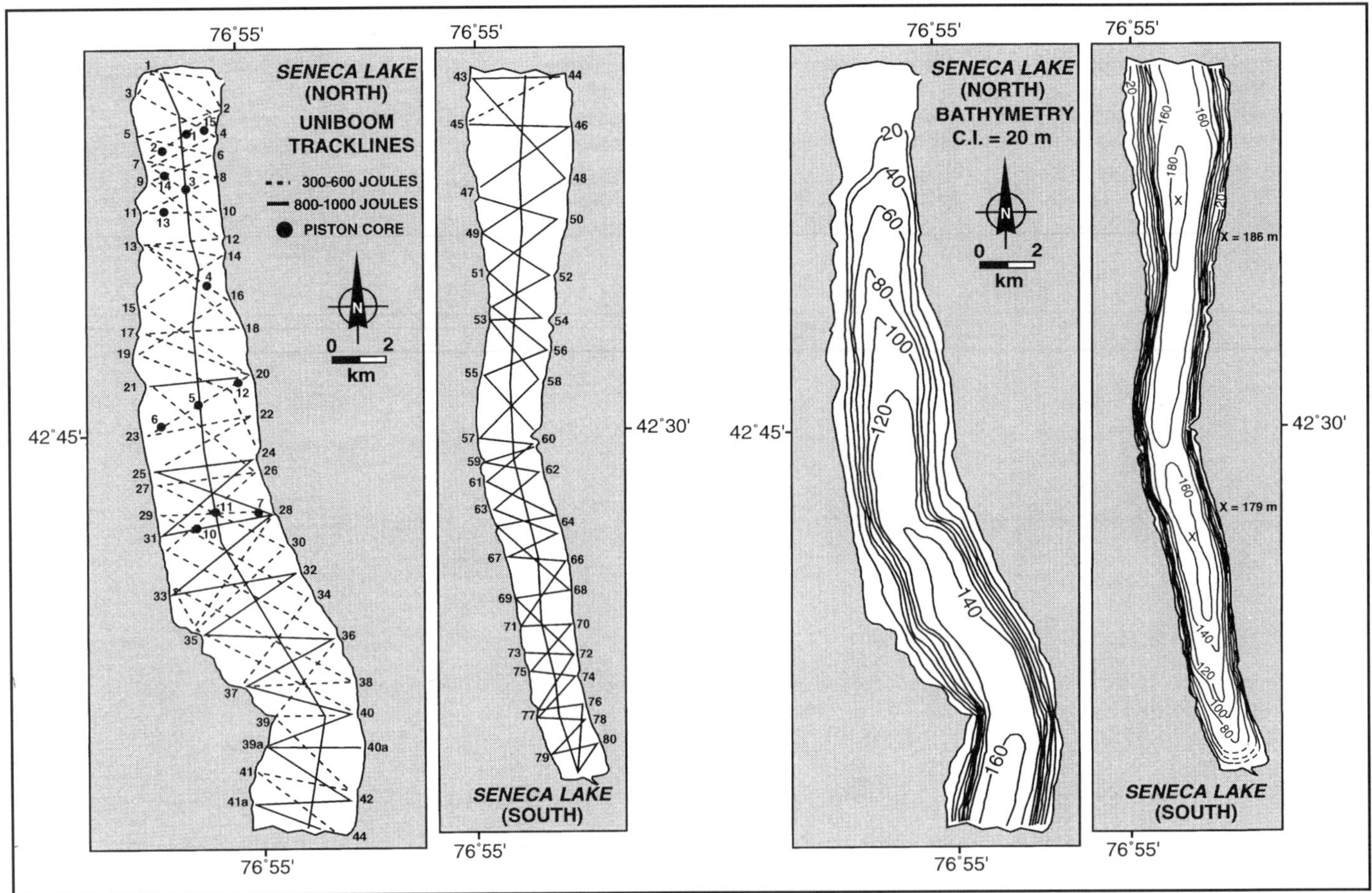

Figure 20. Seismic reflecfion trackline and bathymetric maps for Seneca Lake. Note that for purposes of presentation, Seneca Lake has been divided into northern and southern halves. Key for all Seneca Lake maps: G = Geneva; KLO = Keuka Lake Outlet; WG = Watkins Glen, C.I. = contour interval. Piston core locations also shown on trackline map.

Lake, we interpret these buried ridges of sequence II in Seneca Lake as preserved esker segments.

As in the other Finger Lakes, sequence IV beneath Seneca Lake is characterized by a continuous, high-frequency acoustic facies in which individual reflections can be traced for tens of kilometers along the longitudinal axis of the lake (Figs. 21 and 25). Sequence IV is thickest in the northern half of the lake basin, indicating a northerly source of sediment (Fig. 24). However, sequence IV also thickens significantly in the central portion of the basin near Long Point (Fig. 24), suggesting an additional point source of sediment. Today, the Keuka Lake Outlet flows into Seneca Lake just north of Long Point (Fig. 31). Based on detailed isopach mapping of subunits within sequence IV, Stephens (1986) suggested that the Keuka Lake Outlet previously entered Seneca Lake at Long Point and has since migrated northward. Collectively, these data indicate that sediment entered Seneca Lake during sequence IV, in part via outflow from Keuka Lake, suggesting that Keuka Lake was uncovered by ice.

In order to help groundtruth our seismic stratigraphy of Seneca Lake, we collected 15 piston cores (up to 5 m long) from the northern half of the lake at water depths ranging from 11 to 146 m (Figs. 20 and 26). As in Cayuga Lake, piston core sites were strategically located at seismically defined outcrops, which allowed us to sample sequences IV, V, and VI. Sequence IV was found to consist of pink to gray, organic-lean muds that are distinctly laminated on a centimeter scale. These rhythmites (varves?) of sequence IV are occasionally interrupted by thicker massive mud units, thin graded sand layers, diamicts, and occasional dropstones. Collectively, these core results point to a proglacial lacustrine environment for the deposition of sequence IV.

Depositional sequence V consist of massive brown muds with occasional laminations (Fig. 26). In all the Finger Lakes, sequence V records a drainage reversal and the rapid influx of sediment from the south that is consistent with its largely massive, fine-grained nature in the cores. In contrast, sequence VI was found to consist of two distinct sediment types. In relatively shallow littoral zones of the lake, sequence VI consists of lacustrine marl (relatively high in $CaCO_3$) with abundant fossil mollusks and macroscopic plant fragments (Fig. 26), whereas

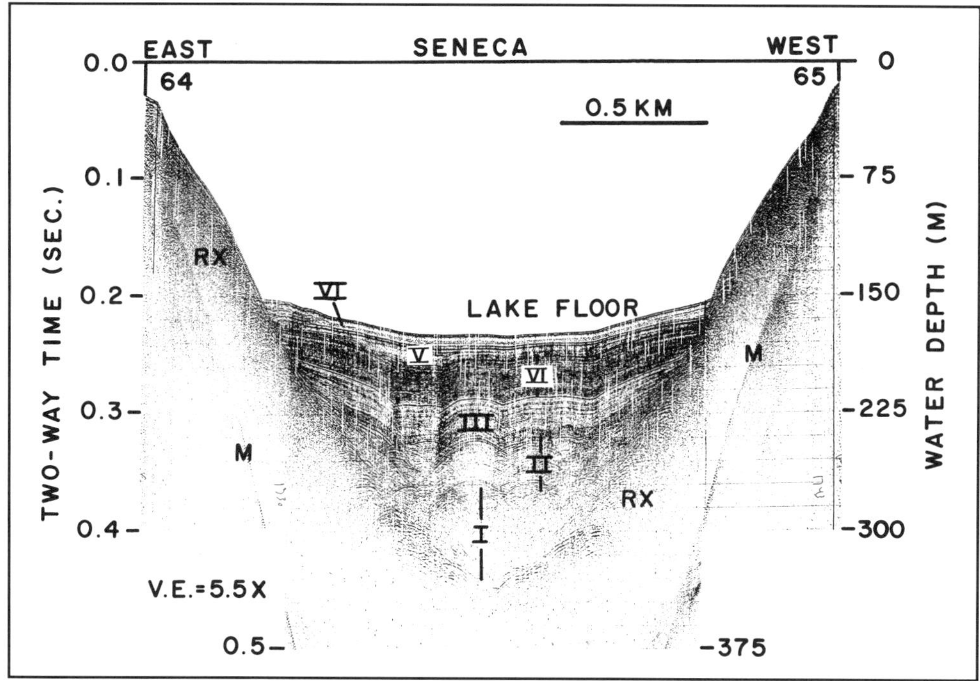

Figure 21. Transverse seismic reflection profile 64-65 from southern Seneca lake (see Fig. 20) illustrating seismic stratigraphic divisions of sediment-fill as well as bedrock reflection (RX), which can be traced continuously across the lake.

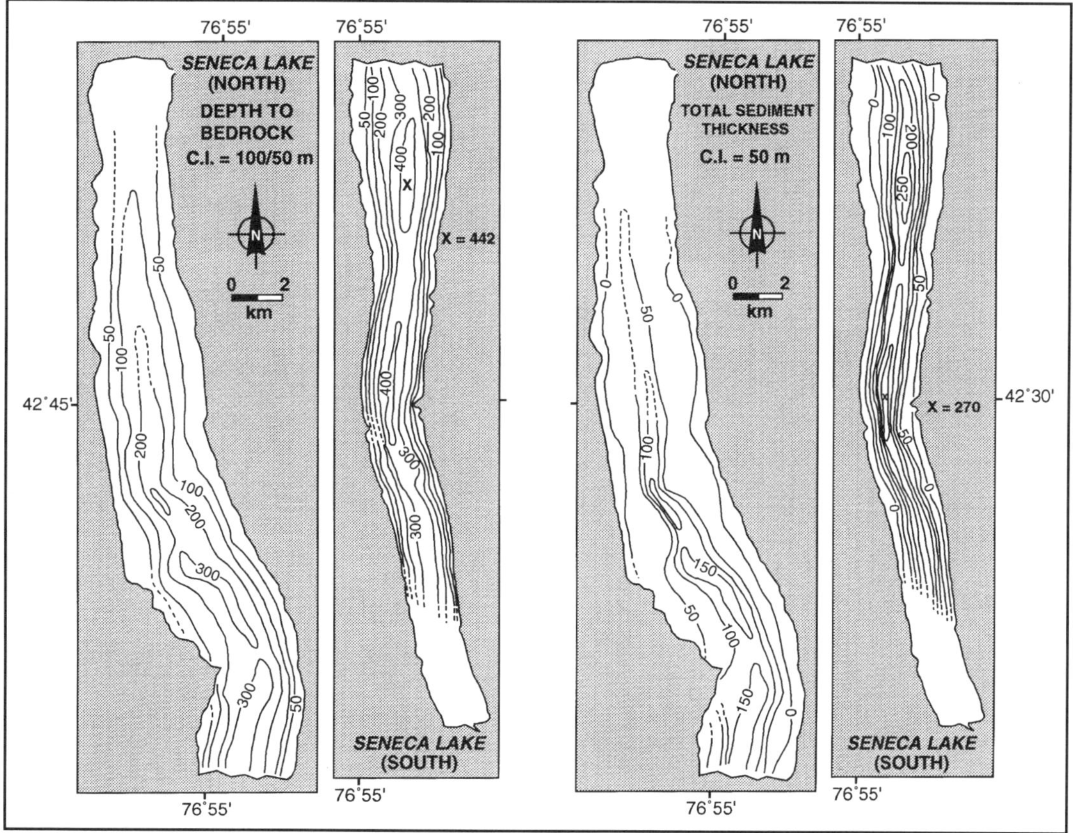

Figure 22. Structure contour map on top of eroded bedrock surface (left) and total sediment thickness isopach map (right) for Seneca Lake. Note closed bedrock basins along axis of Seneca Lake. Data are not available at north end of lake due to acoustic impenetrability (i.e., gas) and at south end of lake due to hummocky lake floor.

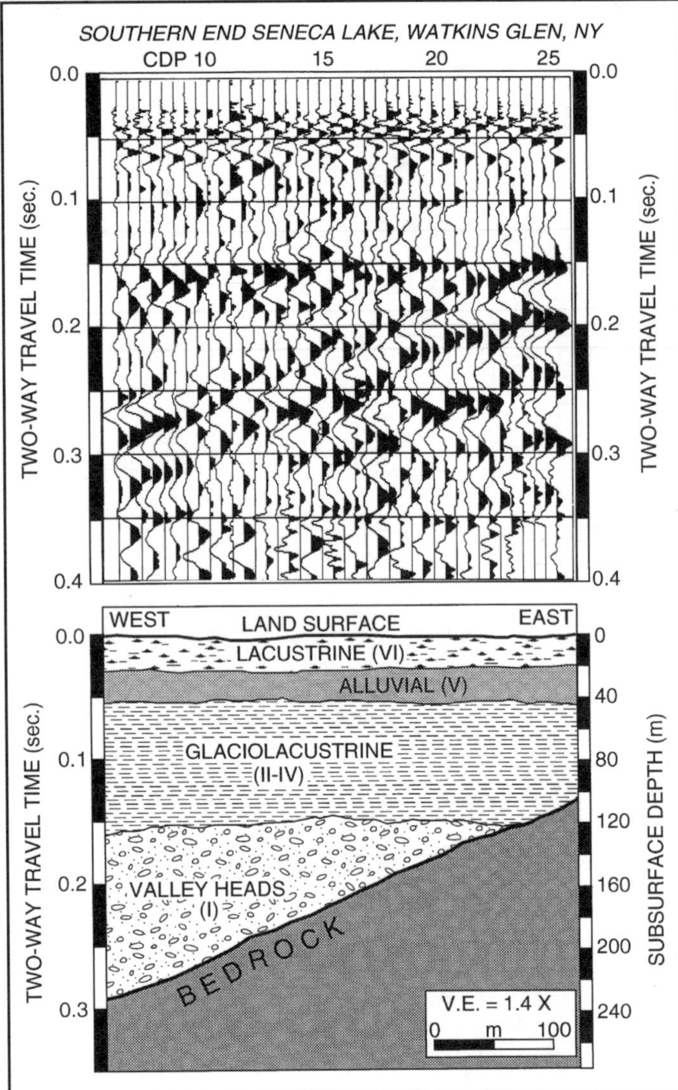

Figure 23. Photograph (top) and line drawing interpretation (bottom) of onland, weight-drop, multichannel seismic reflection profile collected from the eastern half of Seneca Valley 3.5 km south of the lake.

the deep profundal portion of the lake floor is underlain by dark gray to black, laminated organic-rich muds with only occasional fossils (Fig. 26).

An accelerator mass spectrometer (AMS) radiocarbon date on a wood fragment recovered from a subsurface depth of 294 cm in core 8 has an age of 8,320 ± 70 B.P. (ISOTRACE Lab sample TO-1068). Assuming that the top of core 8 is zero years, this date yields an average accumulation rate of about 35 cm/1,000 yr. Projecting this average accumulation rate to the base of sequence VI at core site 8 (4.7 m on seismic line 39-40) results in an overall age *estimate* for sequence VI of 13.4 ka. This age estimate of 13.4 ka agrees well with a radiocarbon age of 13,925 ± 700 yr from just below the sequence VI-V boundary in Seneca Lake reported by King et al. (1983), as well as a recently obtained AMS date on bulk organics of 13,865 ± 100 yr B.P. (Arizona Lab sample AA15245) from just above the VI-V boundary in Seneca Lake. These data are also in very good agreement with a radiocarbon age of 13.6 ± 0.2 ka from just above the sequence VI-V contact in Canandaigua Valley, which has been confirmed via paleomagnetism (see Wellner et al., this volume).

Overall, lithostratigraphic results of our piston coring efforts are consistent with previous bottom sediment studies in Seneca Lake. Woodrow et al. (1969) grouped bottom sediments found in central Seneca lake as either heterogeneous, coarse-grained near-shore facies; fine-grained, brown, cohesive lake slope facies; or fine-grained, black to dark gray deep lake facies. In the northern part of Seneca Lake, Woodrow (1978) also reported relatively shallow-water sandy marl and deep-water black muds separated by a narrow erosional outcrop of pink clays (characterized by "well defined, parallel reflectors") which he interpreted as proglacial lake deposits.

Keuka Lake

Keuka is the most unusual Finger Lake. With its overall Y-shape it is the only Finger Lake with branches (Fig. 1). The two branches of Keuka Lake converge southward, which strongly suggests a southward-directed preglacial drainage system. Keuka Lake also is the only Finger Lake that does not occur in a through valley. All other Finger Lakes drain independently to the north. However, today, outflow from Keuka drains to the east into Seneca Lake via the Keuka Lake Outlet.

Approximately 180 km of seismic reflection profiles have been collected from Keuka Lake (Fig. 27). Our bathymetric map displays yet another unusual feature of Keuka Lake (Fig. 27). Unlike the other Finger Lakes that gradually deepen from north to south, the Keuka Lake floor has a relatively constant water depth of 40 to 50 m along its northwest branch and 30 to 40 m along its northeast branch. Additionally, there are a series of closed bathymetric depressions along much of the lake floor. Maximum water depth (57 m) occurs in one of these axial depressions near the confluence of Keuka's two branches (Fig. 27). However, in transverse profile, Keuka is similar to the other Finger Lakes in being relatively steep-sided and flat-floored.

Data on depth to bedrock beneath Keuka Lake (Fig. 28) document that the shallower, northeastern branch of the lake is actually a hanging valley of the northwest branch. The northwest branch of Keuka Lake has been eroded as much as 193 m below lake level (25 m above sea level), whereas the northeast branch has been eroded only to depths of 150 m below lake level (Fig. 28). Both branches, however, have a series of closed bedrock depressions along their axis. At the confluence of the two branches near Bluff Point, there is a distinct bedrock high or sill where the northeast branch joins the northwest branch (Fig. 29). From the crest of this bedrock sill to the adjacent bedrock depression along the northwest branch there is approximately 125 m of relief (Fig. 28). Correlation of seismically

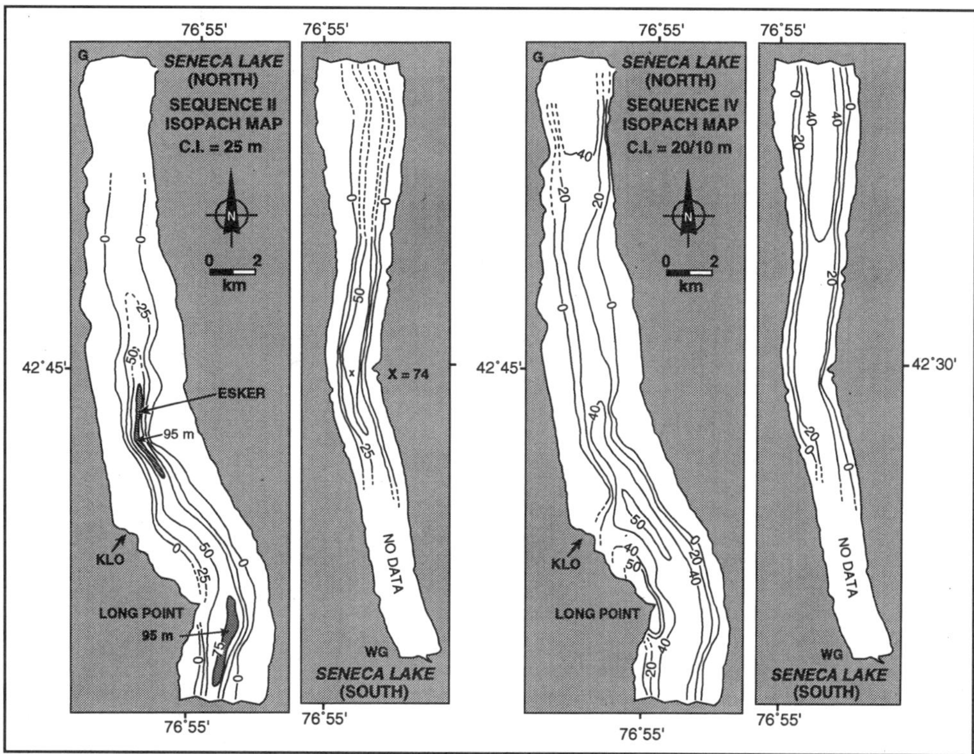

Figure 24. Isopach maps of sequence II (left) and sequence IV (right) in Seneca Lake. Note distribution of sequence II eskers along axial thalweg of northern Seneca Lake and thickening of sequence IV off Long Point.

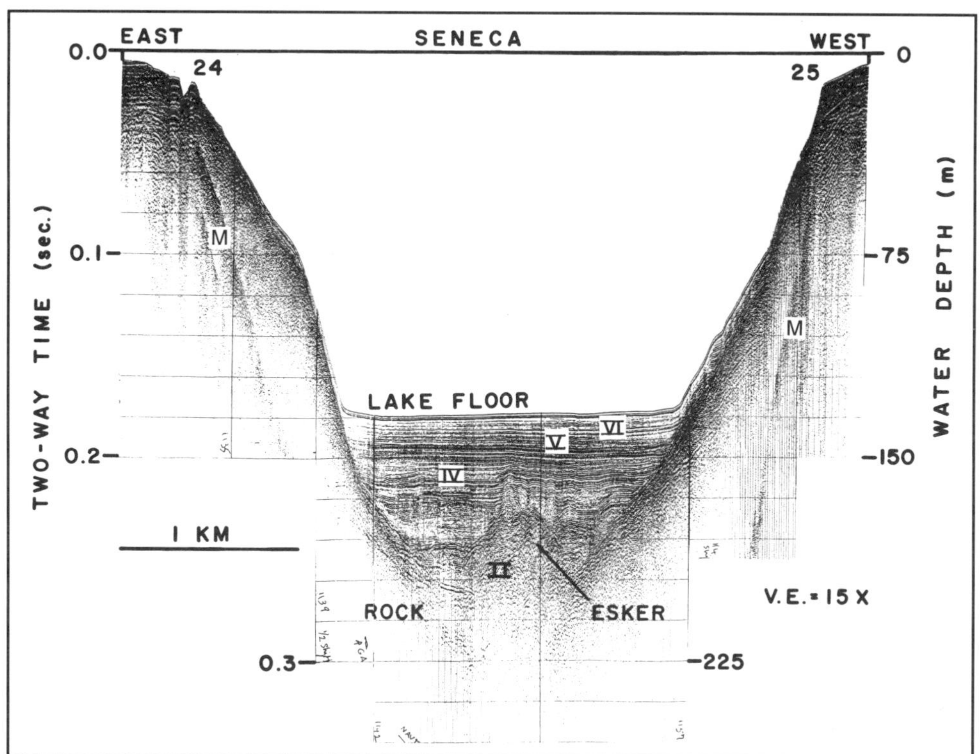

Figure 25. Transverse seismic reflection profile 24-25 from northern Seneca Lake illustrating buried esker in sequence II.

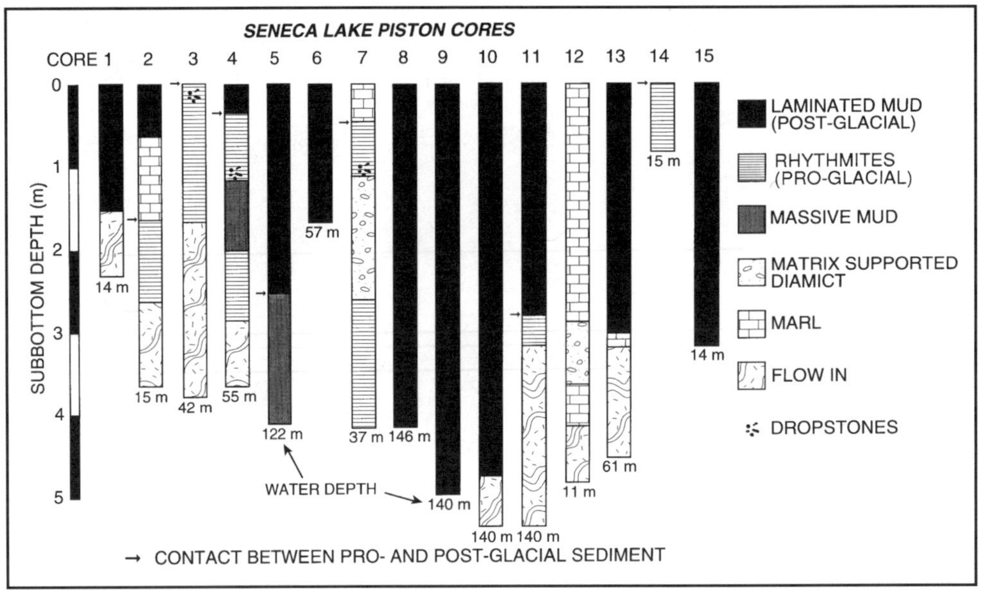

Figure 26. Schematic, lithologic description of piston cores recovered from Seneca Lake (see Fig. 20 for locations).

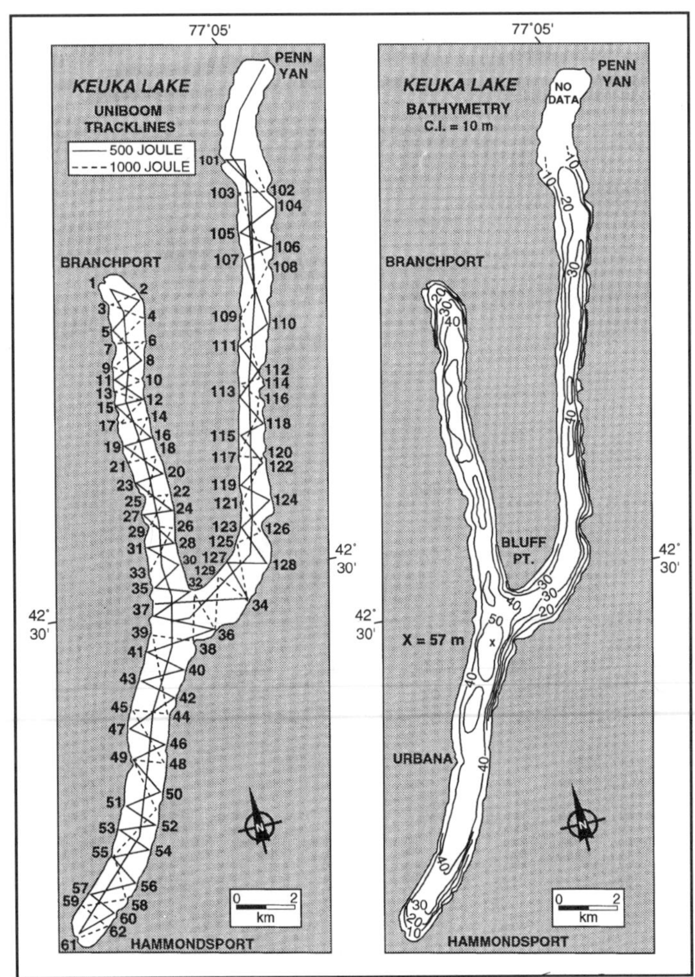

Figure 27. Seismic reflection profile trackline and bathymetric maps from Keuka Lake. Note closed bathymetric depressions along axes of both branches of the lake.

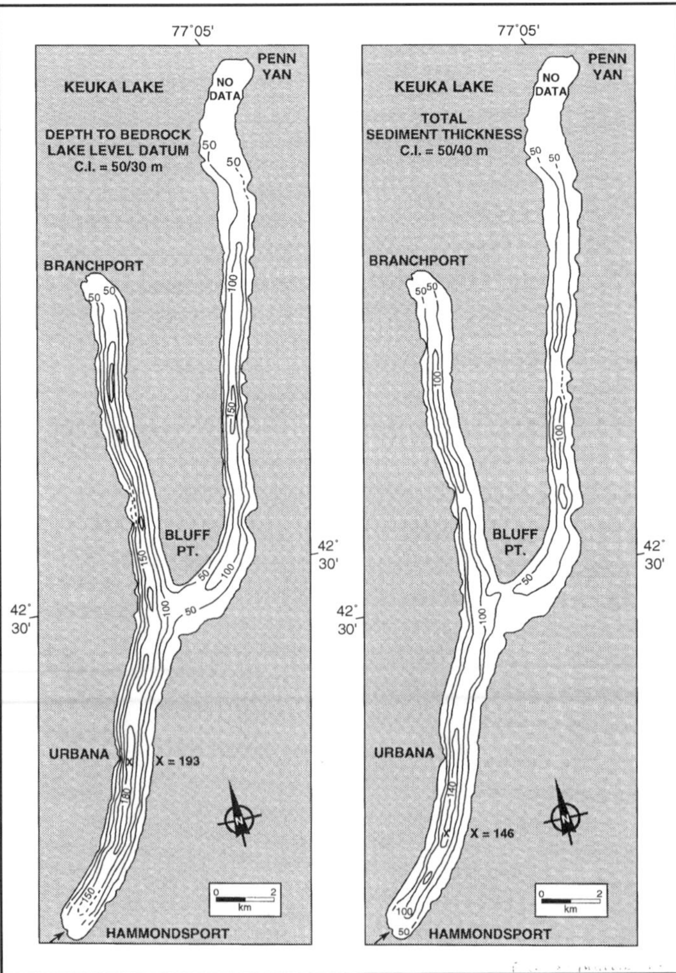

Figure 28. Structure contour map on top of eroded bedrock surface (left) and isopach map of total sediment-fill (right) for Keuka Lake.

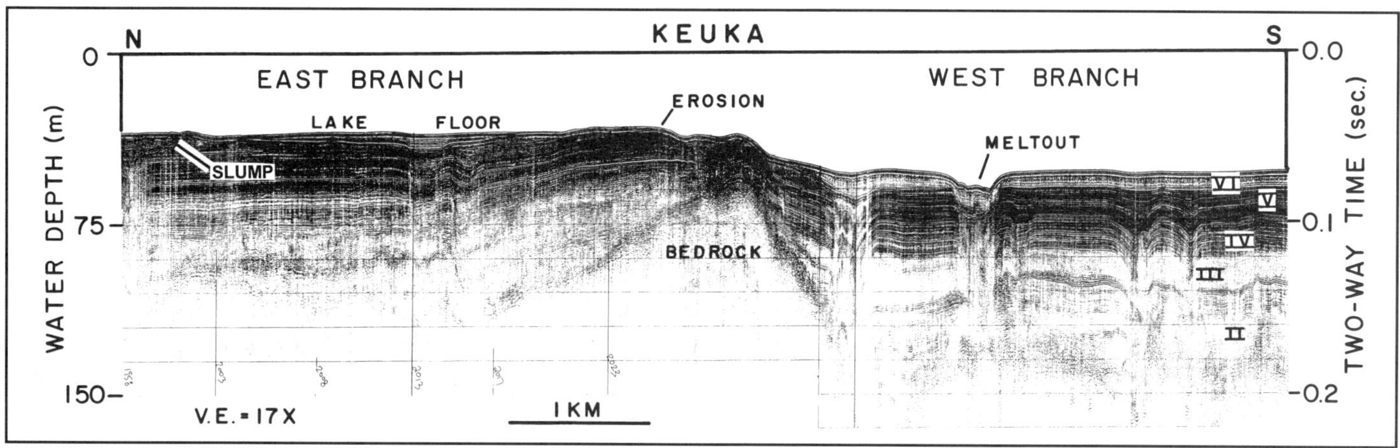

Figure 29. Portion of axial seismic reflection profile from Keuka Lake extending from the northeast to northwest branches of the lake at their confluence near Bluff Point (see Fig. 27). Note bedrock high (sill) where the branches of the lake meet.

defined depositional sequences between the two branches reveals that sequences I through VI occur in the northwest branch, whereas only sequences II through VI are found in the northeast branch. This is reflected in total sediment thickness distributions (Fig. 28). Maximum sediment thickness in the northeast branch is on the order of 100 m, whereas the northwest branch contains (near its southern end) as much as 146 m of total sediment. The six depositional sequences that make up this total sediment fill beneath Keuka Lake are acoustically very similar to those described from the other Finger Lakes.

Reflection profiles that cross bathymetric depressions in Keuka Lake reveal evidence of deformed sediments (Figs. 30 and 31). Depositional sequences III through V have clearly been disturbed via collapse of underlying material in sequence II, whereas depositional sequence VI has passively infilled preexisting lake floor depressions without further deformation. We interpret these bathymetric depressions and underlying deformed sediments as a product of the meltout of buried ice blocks. The stratigraphic relationships displayed on the siesmic profiles argue for the rapid burial of detached ice blocks during sequences III through V, followed by melting prior to the deposition of sequence VI. Based on the distribution of deformed subsurface sediments, we have mapped 16 individual meltout structures along the axes of both branches of Keuka Lake south to the town of Urbana. These meltout structures are likely analogous to "dead-ice sinks" described by Fleisher (1986) from glaciated valleys of the Appalachian Plateau in central New York State. Fleisher (1986) argued that stagnant ice features preferentially develop in non–through valleys (such as Keuka) due to detachment of ice blocks during retreat over preexisting drainage divides. Thus, the abundance of meltout structures beneath Keuka Lake can be understood in terms of its non–through valley morphology, rather than as an indication of regional downwasting (stagnation) of the Laurentide ice sheet during its retreat across the Finger Lakes region. Keuka is the only Finger Lake in which significant numbers of ice meltout structures have been identified.

Canandaigua Lake

Canandaigua is the westernmost of the seven larger eastern Finger Lakes (Fig. 1). Immediately south of Canandaigua Lake is a wetland and well-developed dry lake valley that extends 8 km to the village of Naples where outcrops of the Valley Heads Moraine rise ~200 m above the dry lake floor (Fig. 2).

Approximately 120 km of high-resolution seismic reflection profiles were collected from Canandaigua Lake (Fig. 32). The bathymetry of Canandaigua Lake differs from that of the other Finger Lakes. Instead of being longitudinally asymmetric with maximum water depths in the southern half of the lake basin, Canandaigua is a longitudinally symmetrical lake basin with a maximum water depth of 84 m near the midpoint of the basin (Fig. 32). In transverse section, Canandaigua is similar to the other Finger Lakes in that it is relatively steep-sided and flat-floored.

Glacial erosion of bedrock beneath Canandaigua Lake extends as much as 261 m below lake level in a series of closed axial depressions (Fig. 33). Considering that Canandaigua's modern lake level is at an elevation of 210 m, bedrock beneath the lake has been eroded as much as 51 m *below* sea level. Similar to most of the other Finger Lakes, bedrock erosion deepens from north to south, reaching a maximum in the southern third of the lake basin (Fig. 33). Thus, Canandaigua is unusual in that maximum water depth and maximum bedrock erosion are not coincident, suggesting that modern water depths have been shaped as much (or more so) by sediment infill as they have by bedrock erosion. Total sediment fill, however, does follow trends in the underlying eroded bedrock surface. Total sediments are relatively thin (<50 m) in the northern end of the basin but thicken southward, reaching a maximum of 202 m in the southern third of the lake, which is coincident with maximum bedrock erosion (Fig. 33). All six acoustically defined depositional sequences in the Finger Lakes region occur beneath Canandaigua Lake. However, sequences III and IV are particularly well developed and are highlighted here.

Depositional sequence III is an acoustically transparent (reflection-free) unit (Fig. 34) up to 53 m thick that is restricted to the central portion of the lake basin (Fig. 35). The lack of reflections from within sequence III indicates a lack of significant acoustic impedance (density × velocity) contrasts, which implies that sequence III is a massive (homogeneous) deposit. Although sequence III is the only Finger Lake depositional sequence that has not been sampled, the complete penetration of this sequence by the Uniboom acoustic pulse, coupled with the recording of high-amplitude reflections from beneath it,

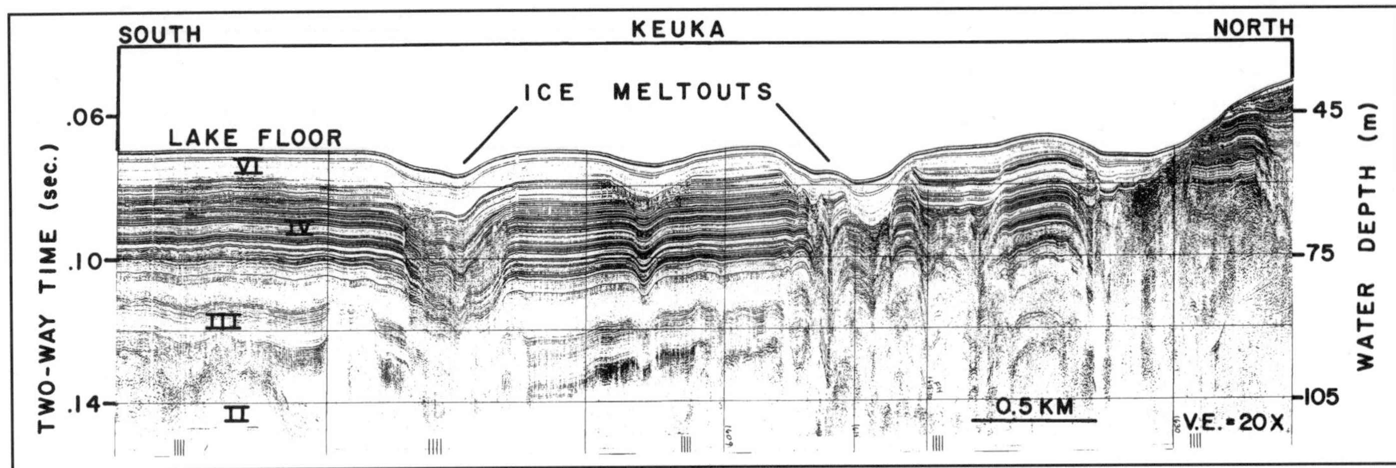

Figure 30. North end of axial seismic reflection profile from the northwest branch of Keuka Lake (see Fig. 27) illustrating ice meltout structures. Note that bathymetric depressions are underlain by downdropped and deformed reflections.

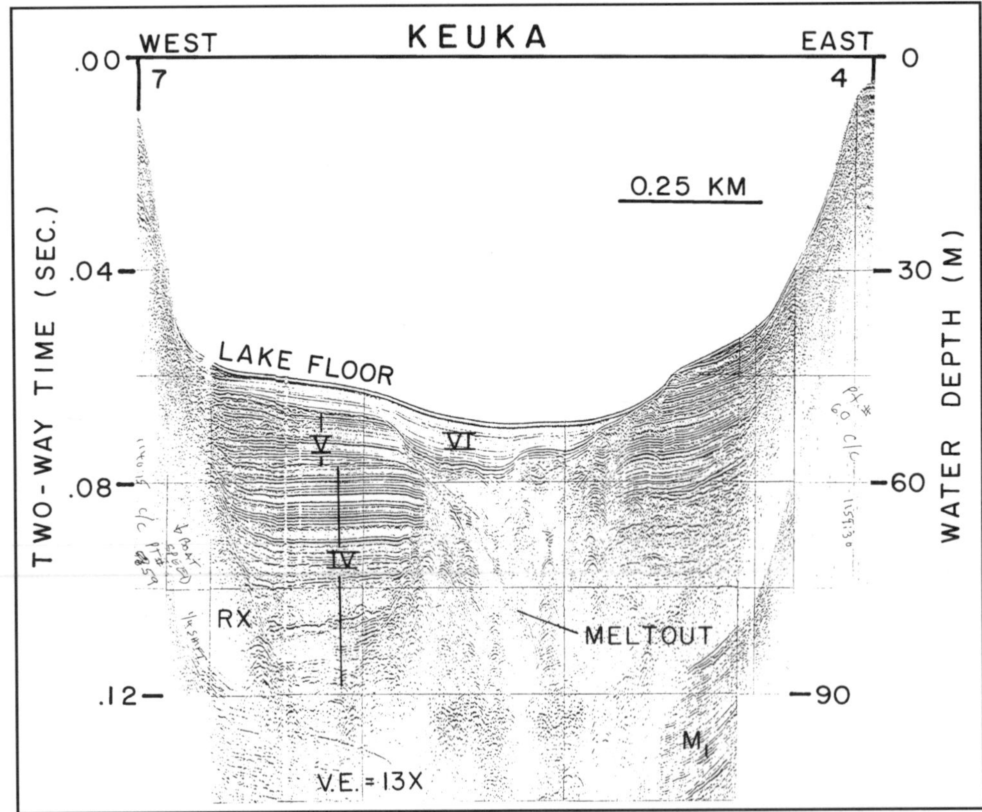

Figure 31. Transverse seismic reflection profile 4-7 (see Fig. 27) from northern end of the northwest branch of Keuka Lake illustrating an ice meltout structure. Note coherent reflections to either side of bathymetric depression, which is underlain by chaotic reflections with numerous diffractions. Also note undeformed nature of sequence VI reflections.

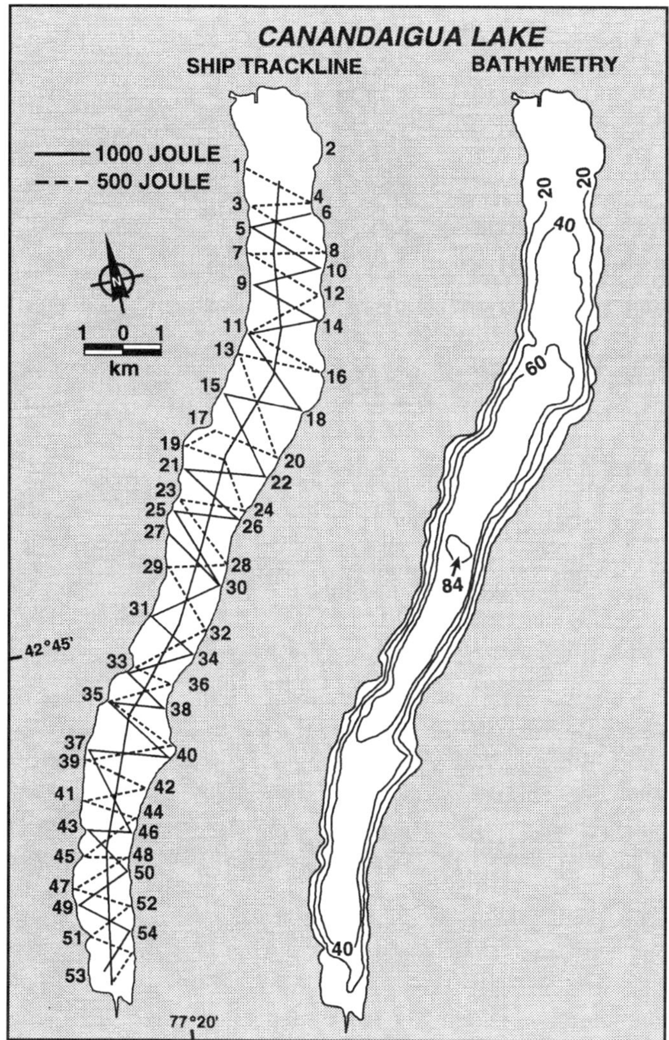

Figure 32. Seismic reflection profile trackline and bathymetric maps of Canandaigua Lake.

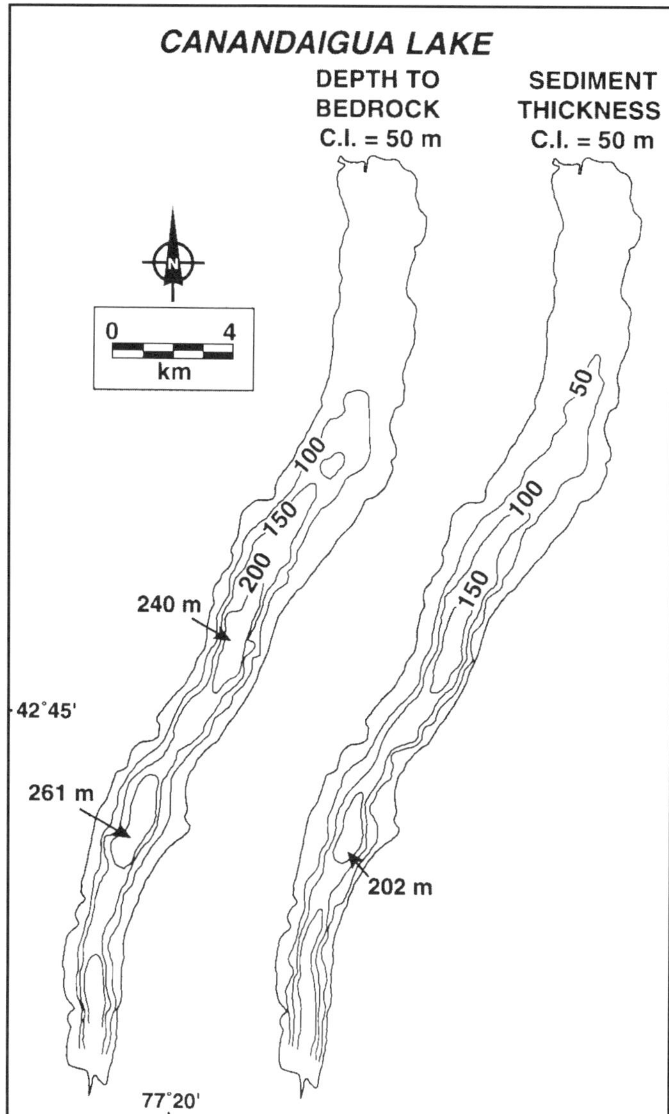

Figure 33. Depth to bedrock and total sediment thickness maps of Canandaigua Lake.

argues that sequence III consists of very fine grained sediments, sediments that may be structureless due to very high rates of accumulation.

Depositional sequence IV beneath Canandaigua Lake is similar to the other Finger Lakes in that it consists of high-frequency, continuous reflections that can be traced for kilometers along the longitudinal axis of the lake (Fig. 34). However, unlike the other Finger Lakes, sequence IV is thickest (43 m) in the southern half of the lake basin (Fig. 35), suggesting that it was derived largely from the *south.* In contrast, sequence IV in the other Finger Lakes appears to be northerly derived glaciolacustrine (proglacial) deposits. In this regard Canandaigua Lake may be more akin to Otsego Lake (east of the Finger Lakes near Cooperstown, New York), where proglacial sediments were transported to the south end of the lake basin via ice-free tributary streams (Fleisher et al., 1992). Alternatively, drainage reversal, produced by a lowering of proglacial lake levels due to ice withdrawal to the north, may have begun sooner in the Canandaigua Lake basin than it did in the other Finger Lakes.

Following analysis of seismic reflection data from all of the Finger Lakes, we had hoped to recover a long drillcore from within one of the lakes. However, funding limitations forced us to consider an onland site. We selected a drillsite 3 km south of Canandaigua Lake for the following reasons: (1) there was an accessible dry lake valley south of the lake, (2) lacustrine-based seismic data from the south end of the lake suggested that all depositional sequences (with the exception of III) found within the lake projected southward beneath the dry valley (Fig. 36) and, (3) onland multichannel seismic reflection data indicated that there was a thick (~155 m) sediment-fill beneath the dry lake valley (Fig. 37). Two drillcores were recovered from the dry valley south of Canandaigua Lake: drillcore 1, 118 m long,

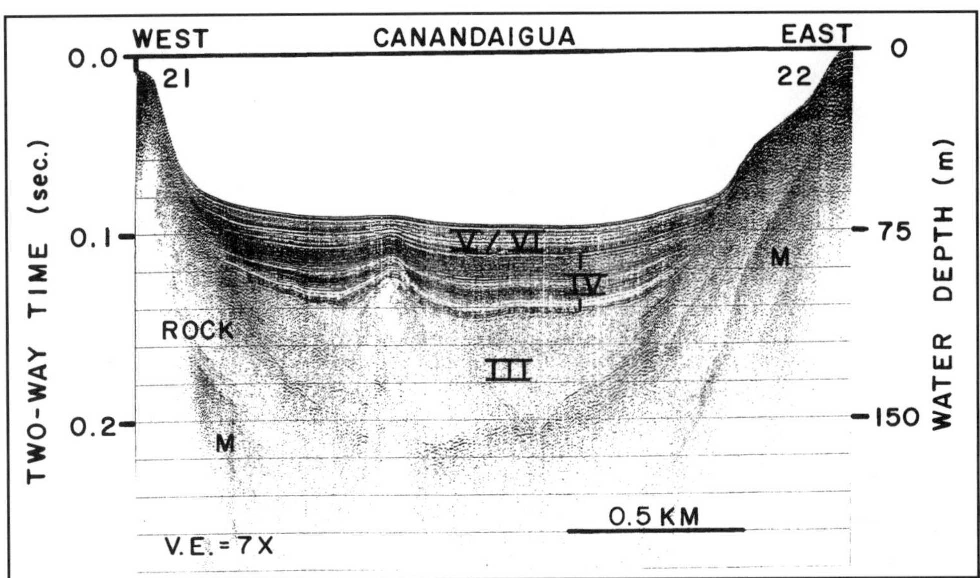

Figure 34. Transverse seismic reflection profile and line drawing interpretation of profile 21-22 (see Fig. 32) from Canandaigua Lake illustrating depositional sequences III through VI.

and drillcore 2, 13 m long (Fig. 37). Here, only first-order results from drillcore 1, as they pertain to the regional stratigraphy of the Finger Lakes, are presented. Details, including downhole geophysics, of drillcore 1 can be found in Wellner et al. (this volume), as well as in Nobes and Schneider (this volume); details of drillcore 2 can be found in Wellner and Dwyer (this volume).

Drillcore 1 penetrated 118 m of unconsolidated sediment without reaching bedrock. Four first-order stratigraphic units were sampled. (1) A *basal boulder gravel* was recovered between 116 and 118 m subsurface, with boulders larger than the 20-cm-diameter drillbit, which effectively terminated drilling. This basal boulder gravel is interpreted as a subsurface equivalent of nearby outcrops of Valley Heads Moraine (Fig. 36), which was deposited ~14.4 ka (Muller and Calkin, 1993). It is equivalent to seismic sequence I beneath the lakes. (2) Between 26 and 116 m subsurface, drillcore 1 penetrated a thick (90 m) *glaciolacustrine unit* of dominantly gray mud with interbeds of coarser material (Fig. 37). The base of this unit (95 to 116 m subsurface) is a massive clay devoid of dropstones overlain by an 11-m-thick (84 to 95 m subsurface) subunit of clayey gravel. This in turn is overlain by a 20-m-thick (64 to 84 m subsurface) massive clay with numerous gravel-sized dropstones. The upper 38 m (26 to 64 m subsurface) of this glaciolacustrine unit is dominated by rhythmically bedded sands and mud in which the bed spacing and number of dropstones decreases up-section. The lower massive deposits of this glaciolacustrine unit are correlated with seismic sequence II beneath the lake, whereas the upper rhythmically bedded deposits are correlated with depositional sequence IV. (3) Between 12 and 26 m subsurface is a coarse *alluvial unit.* This alluvium consists of coarsening-upward sand to gravel from nearby prograding alluvial fan complexes. These fans represent the outpouring of coarse sediment from local glens as proglacial lake levels dropped abruptly and a drainage reversal occurred (depositional sequence V). Today, this alluvial sand and gravel unit contains artesian water. (4) The upper 12 m of drillcore 1 consist of interbedded peat, fossiliferous marl, and unfossiliferous gray clay capped by soil. The *lacustrine sediments* in this unit probably represent reflooding of the valley due to differential isostatic rebound following ice retreat (Tarr, 1904) and are correlated to depositional sequence VI beneath Canandaigua Lake. A radiocarbon date from an in situ peat layer directly on top of the alluvial gravels yielded an age of 13,650 ± 210 yr. Thus, the vast majority of the unconsolidated sediment beneath Canandaigua Valley was deposited rapidly between ~14.4 (Valley Heads) and 13.6 ± 0.2 ka.

Western Finger Lakes

The western Finger Lakes (Monnett, 1924) consists of four small lakes: Conesus, Hemlock, Canadice, and Honeoye (Fig. 1). Conesus and Honeoye are shallow (Table 1) eutrophic lakes (Bloomfield, 1978) that are acoustically impenetrable. Usable seismic reflection data were obtained only from Hemlock and Canadice Lakes, which are controlled water resources for the city of Rochester (Schaffner and Oglesby, 1978).

Approximately 30 km of seismic reflection profile data have been collected from Hemlock Lake and 14 km from Can-

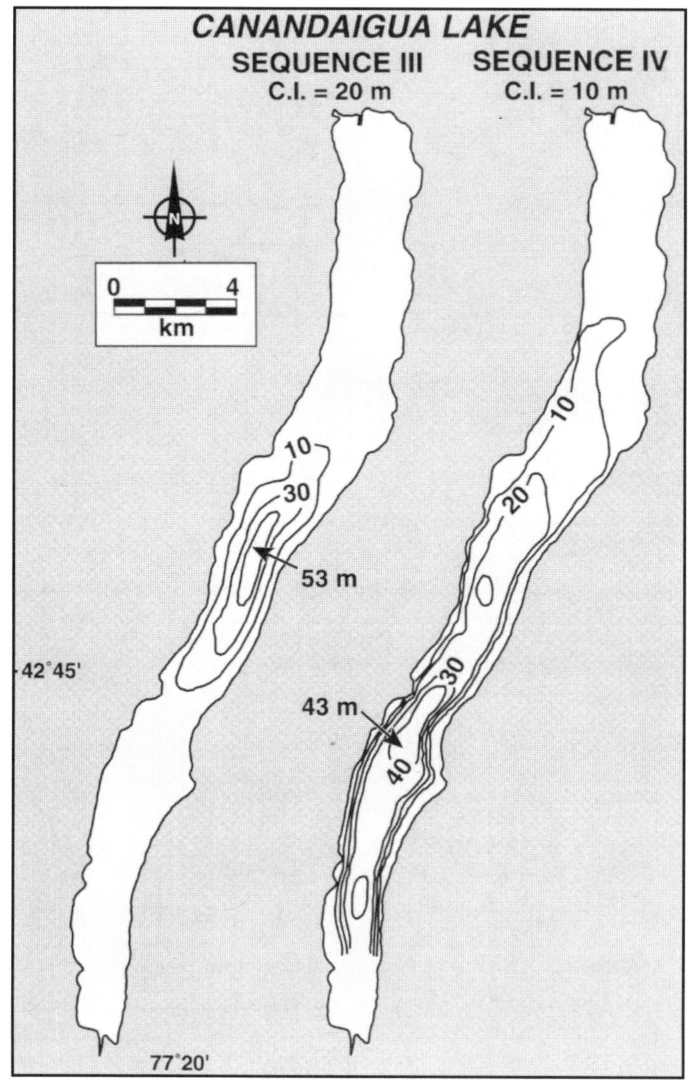

Figure 35. Isopach maps of depositional sequences III and IV beneath Canandaigua Lake.

adice Lake (Fig. 38). Transverse profiles (Fig. 39) reveal that these two western Finger Lakes are relatively steep-sided, flat-floored, deeply scoured, and infilled with thick unconsolidated sediments. Bathymetrically, Hemlock is similar to most other Finger Lakes in that it is longitudinally asymmetric, with maximum water depth (29 m) occurring in the southern half of the lake basin (Fig. 40). Canadice, however, is more symmetrical and bowl-shaped longitudinally with maximum water depths (27 m) found in the northern half of the lake basin (Fig. 41). Hemlock Lake has been eroded as much as 173 m below lake level and infilled with up to 149 m of sediment (Fig. 40). Total bedrock relief for Hemlock Valley, as measured from the top of adjacent divides to maximum bedrock erosion beneath the lake, is on the order of 500 m.

Compared to Hemlock Lake, Canadice Lake has not been as deeply scoured. Maximum depth to bedrock beneath Canadice Lake is only 94 m, which occurs near the midpoint of the basin (Fig. 41). Total sediment fill (up to 68 m) beneath Canadice is also less than in nearby Hemlock Lake. These data define Canadice as a hanging tributary valley to the more deeply incised Hemlock valley. As such, Hemlock and Canadice are, collectively, similar to Keuka Lake where its northeast branch is a hanging tributary valley to its more deeply scoured northwest branch.

All six Finger Lake depositional sequences are present beneath Hemlock Lake, which are nicely displayed on the axial profile from Hemlock (Fig. 42). Sequence I is restricted to the southern half of the basin and thickens southward toward outcrops of the Valley Heads Moraine (Fig. 43). Sequence II accounts for nearly half (71 m) of the 149 m of total sediment-fill beneath Hemlock Lake (Fig. 43). As in the other Finger Lakes, sequence II exhibits a strong acoustic facies change from chaotic, high-amplitude reflections in the north to transparent (reflection-free) in the south (Figs. 39 and 42). In the northern half of the lake, sequences II through V display localized evidence of deformation (collapse), suggesting that ice blocks were detached and buried during sequence II. However,

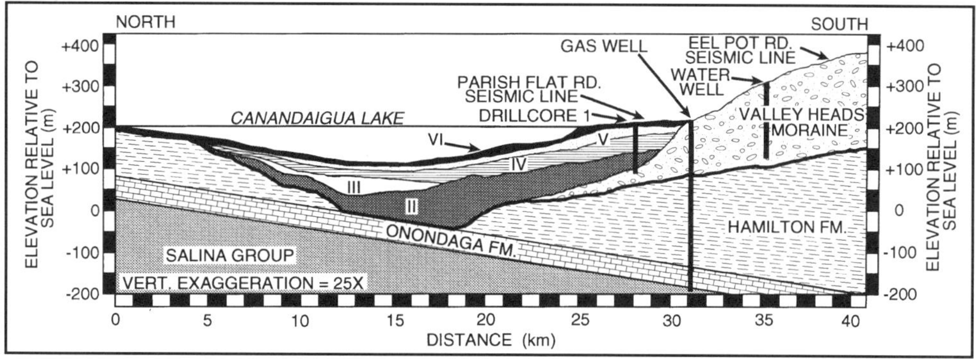

Figure 36. Composite schematic of longitudinal stratigraphy beneath Canandaigua Lake and dry valley based on all available seismic reflection and drillhole data. Note extension of depositional sequences (with the exception of III) beneath the lake southward to drillcore 1.

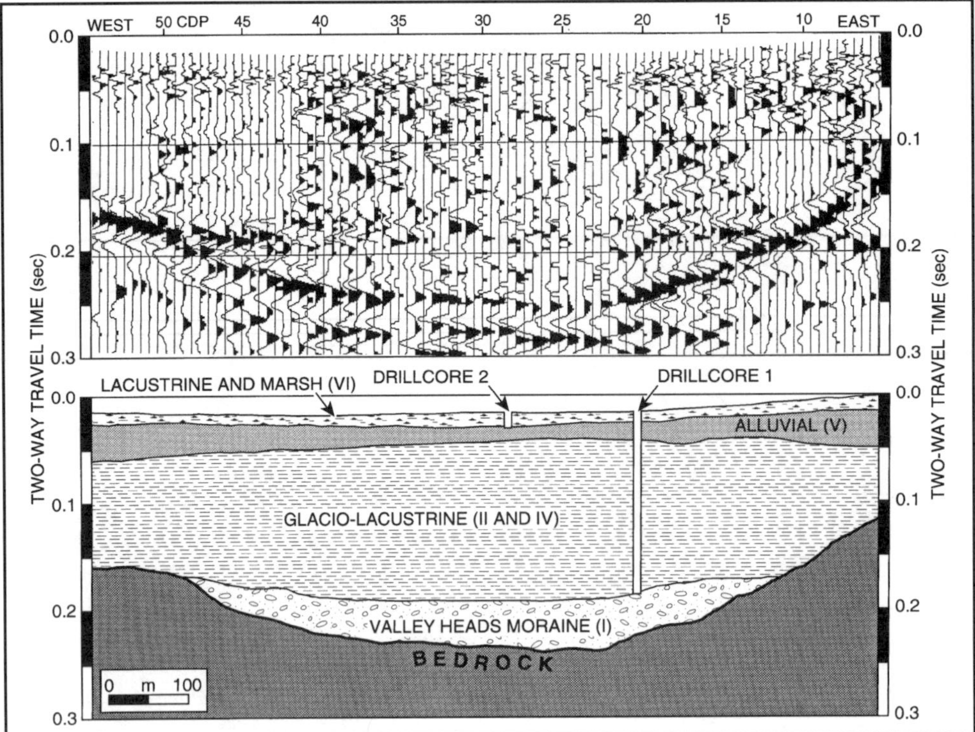

Figure 37. Photograph (top) and line drawing interpretation (bottom) of onland, weight-drop, multichannel seismic reflection profile collected across Parrish Flat Road 3 km south of Canandaigua Lake. Note locations of drillcores and generalized subsurface stratigraphy.

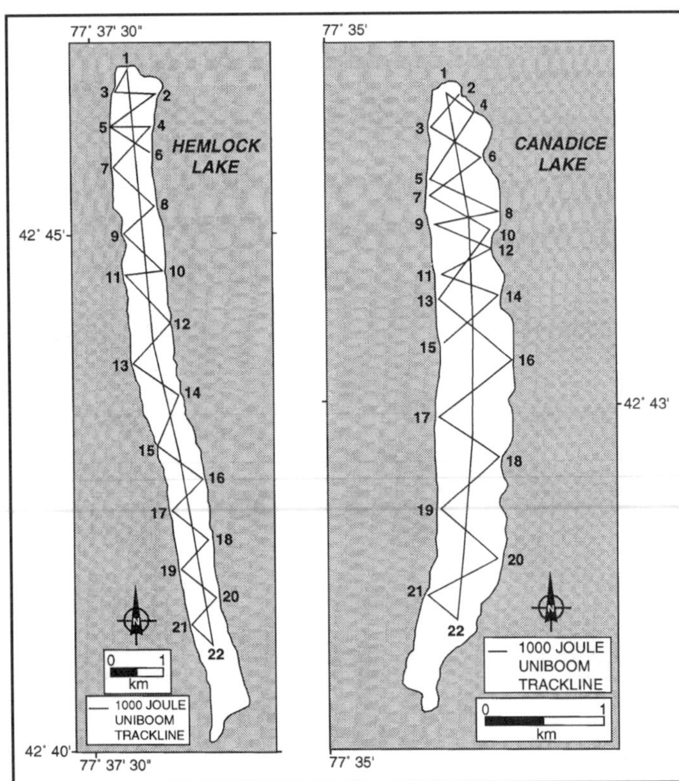

Figure 38. Seismic reflection profile trackline maps of Hemlock and Canadice lakes.

unlike the meltouts in Keuka Lake, those in Hemlock are restricted to the north side of a major step-down in the erosional bedrock surface (Fig. 42), suggesting that the ice was pinned there with either a floating or calving margin.

Sequence III (Fig. 42) is a relatively thin (maximum, 14 m thick), low-amplitude depositional unit restricted to the southern half of the Hemlock Lake basin (Fig. 43). This sequence is found only south of the meltout structures in the lake, suggesting that the ice margin may have temporally lifted from its pinning during the deposition of sequence III.

Sequence IV is characterized by even, parallel, high-frequency reflections that grade from high amplitude in the northern part of Hemlock Lake to low amplitude in the south (Fig. 42). In addition, sequence IV thickens to the north where it is up to 50 m thick (Fig. 44). Collectively, these data suggest a northerly source of sediment with an ice margin to the north of the lake.

Sequence V beneath Hemlock Lake clearly defines a major reversal of drainage and sediment input. Prior to sequence V sediment entered Hemlock Lake from the north with overflow to the south. However, sequence V displays about a five-fold thickening to the south (Fig. 44), indicating that at this time sediment began entering the lake basin from the south, as it does today. As in the other Finger Lakes, this drainage reversal is interpreted as a response to an abrupt lowering of proglacial lake levels and the establishment of modern-day drainage to the north.

Sequence VI in Hemlock Lake is thickest in the deepest,

central part of the lake, strongly suggesting the focusing of "modern" sediments to the deeper parts of the basin (Fig. 44). Maximum sediment thickness for sequence VI is 15 m and occurs over a now-buried ice meltout structure (Figs. 42 and 44), indicating that sequence VI has infilled paleolake-floor topography. Proctor (1978) collected a series of short (<1 m) piston cores from the western Finger Lakes including Hemlock Lake. Based on Cs-137 and Pb-210 data, she calculated sediment accumulation rates for sequence VI of 80 to 350 cm/1,000 yr.

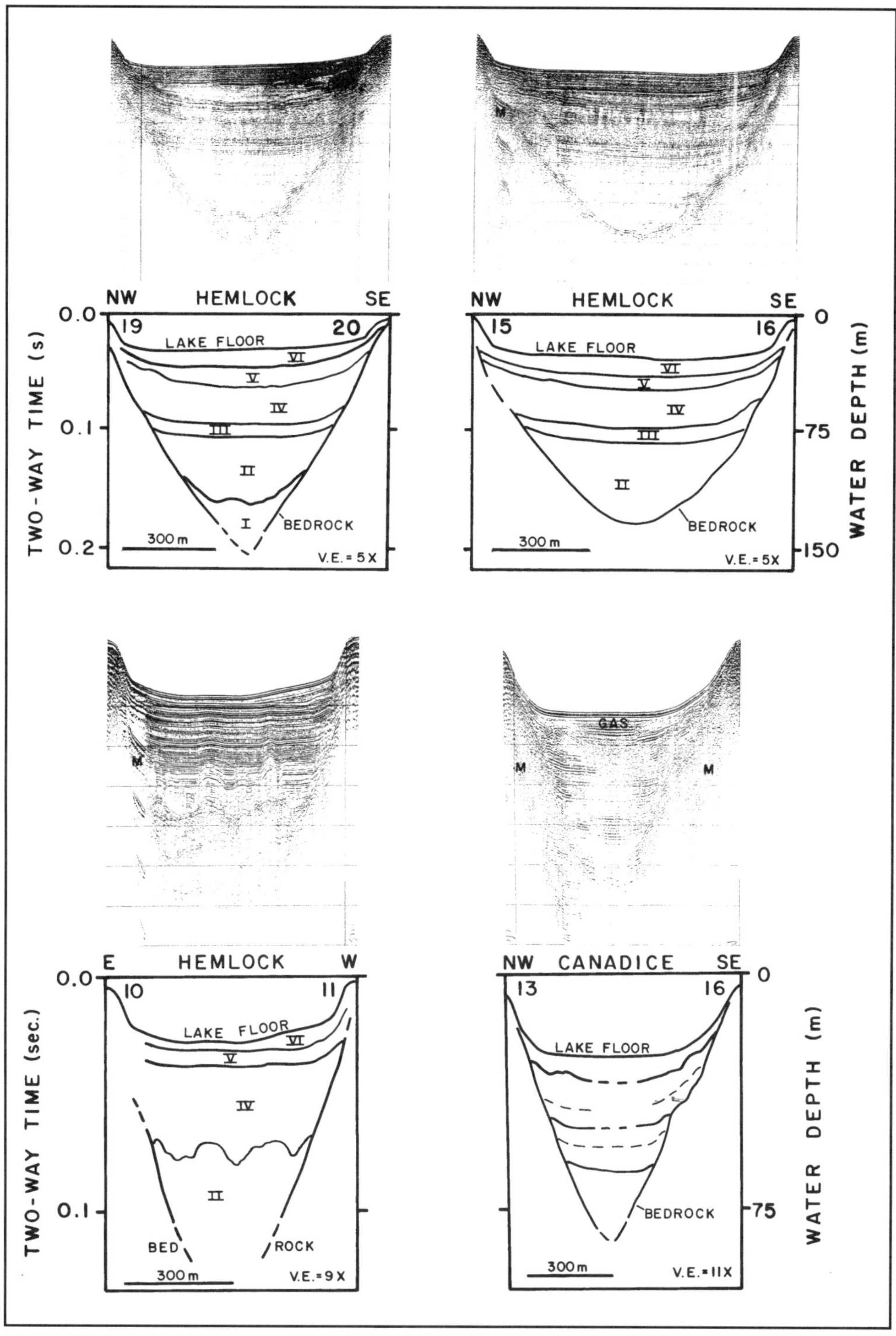

Figure 39. Selected transverse seismic reflection profiles and line drawing interpretations from Hemlock and Canadice lakes illustrating bedrock erosion surface and seismic stratigraphic nature of the sediment-fill. See Figure 38 for locations of profiles.

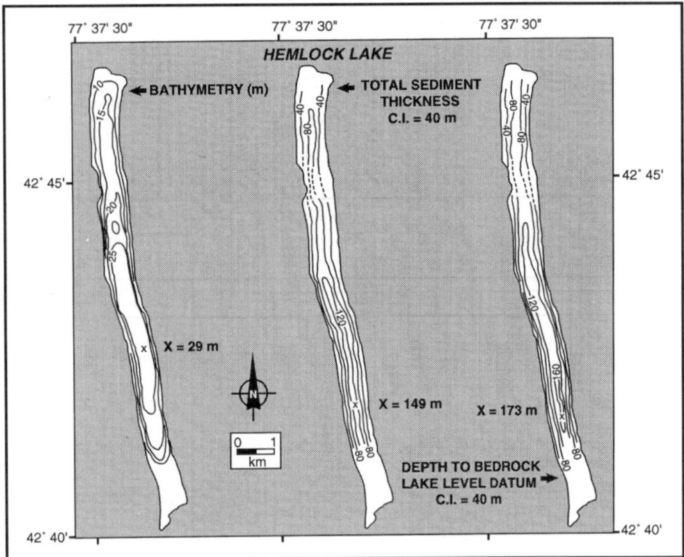

Figure 40. Bathyrnetric (left), total sediment thickness (center), and depth to bedrock (right) maps for Hemlock Lake.

DISCUSSION

Sediment infill history

Our seismic reflection data, coupled with drillcore results and integrated with regional information, document the extensive erosion (as much as 306 m *below* sea level) of the Finger Lake basins and their infill by an unusually thick (up to 270 m) sediment section. Stratigraphic relationships (onlap to Valley Heads) indicate that this thick sediment-fill was deposited during the last (late Wisconsin) glacial cycle. However, this does not preclude the possibility that erosion of the Finger Lake troughs has been the result of multiple glaciations (Coates, 1974; Bloom, 1984) during the past ~2.5 m.y. Any pre–late Wisconsin glacial materials (such as those at Fernbank in Cayuga Valley) that may have existed in the Finger Lake basins must have been removed prior to deposition of late Wisconsin sediments.

The infill history and general processes of deglaciation for the Finger Lakes can be reconstructed and inferred from the six seismically defined depositional sequences found beneath the lakes (Fig. 45). Each Finger Lake displays a similar vertical

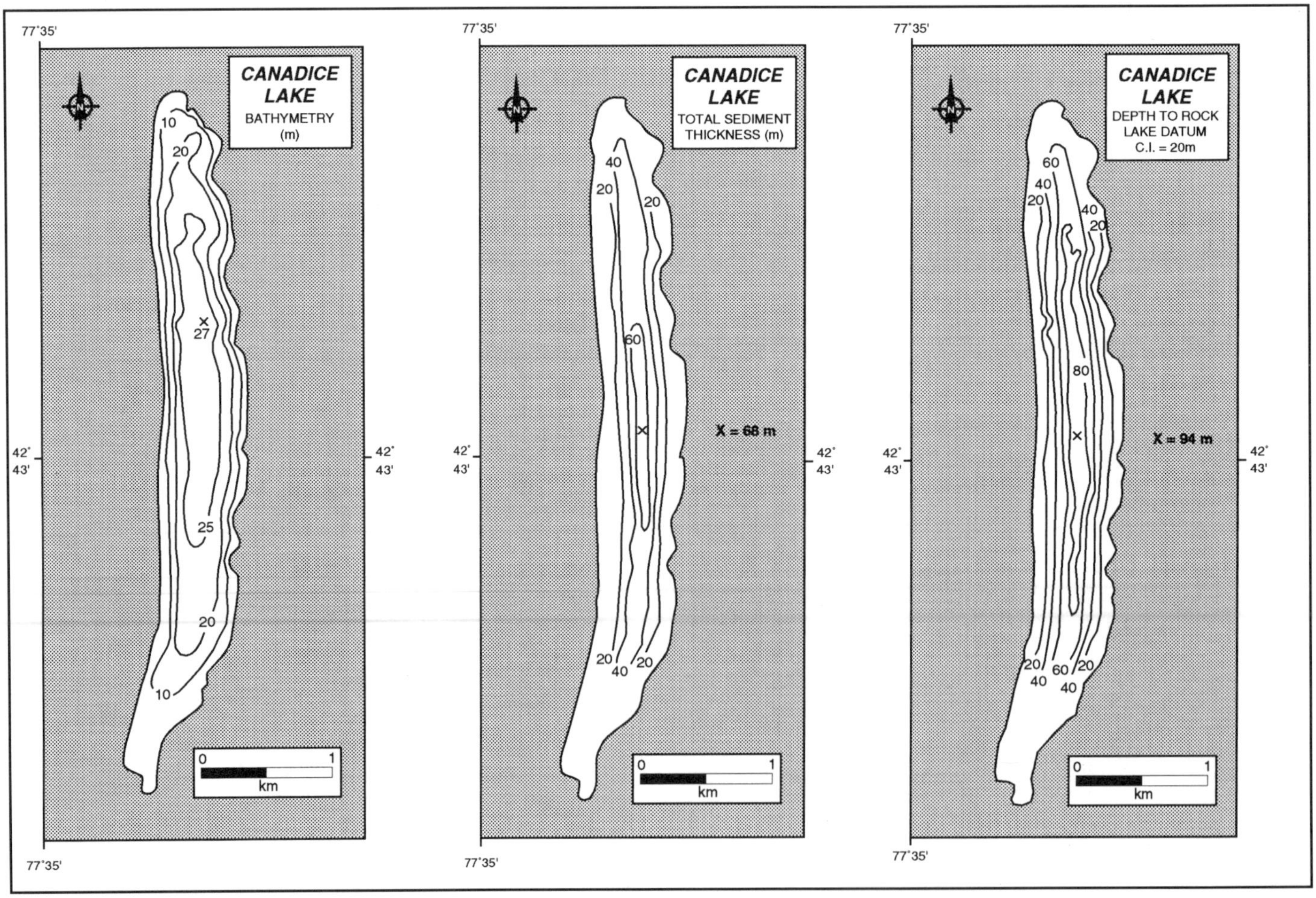

Figure 41. Bathyrnetric (left), total sediment thickness (center), and depth to bedrock (right) maps for Canadice Lake.

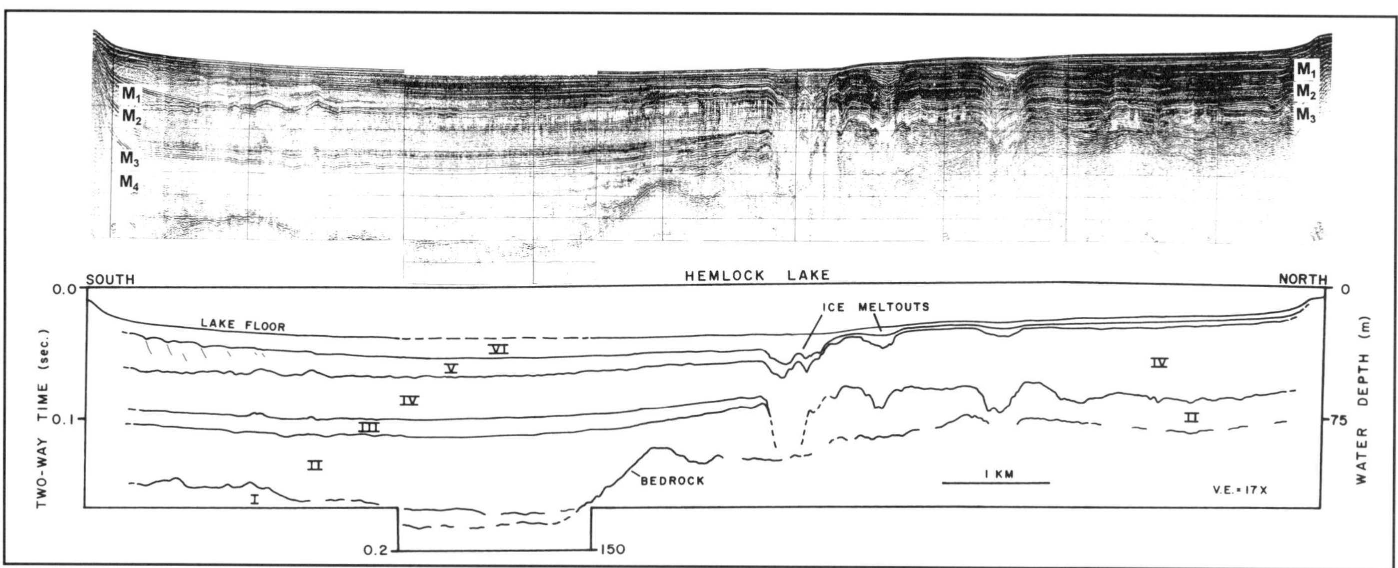

Figure 42. Entire axial seismic reflection profile from Hemlock Lake. Note distributions and facies of depositional sequences, bedrock "step," and ice meltout structures.

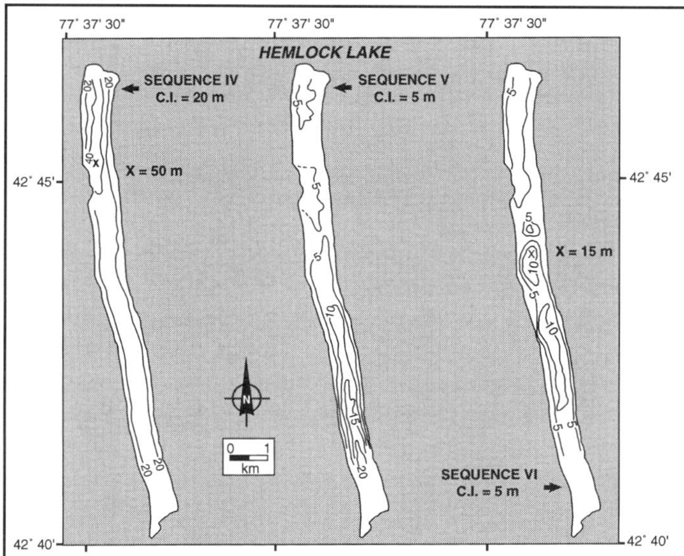

Figure 43. Isopach maps of depositional sequences I through III beneath Hemlock Lake.

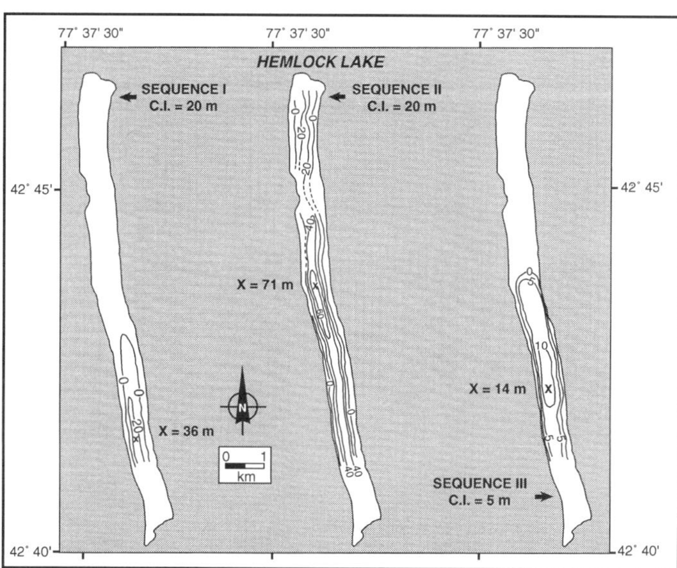

Figure 44. Isopach maps of depositional sequences IV through VI beneath Hemlock Lake.

succession of sequences. However, it does not appear as though the sequences were deposited synchronously in the different lake basins. Reconstruction of ice margin positions (Fig. 46) during deglaciation of the eastern Finger Lakes region, based on the surficial map of Muller and Cadwell (1986), integrated with our results, suggests the Keuka Lake was the first Finger Lake to be uncovered by ice and that the Cayuga Lake basin (lowest lake level elevation) contained ice the longest. Thus, despite the similarity in the vertical succession of depositional sequences in each Finger Lakes, their timing may have been diachronous, at least within confines of the duration of regional deglaciation (<1 ka).

Depositional sequence I beneath the Finger Lakes is equivalent to Valley Heads deposits laid down ~14.4 ka (Muller and Calkin, 1993). The Valley Heads are thick (locally >200 m), coarse-grained deposits of mostly water-laid sand and gravel mapped as kame moraines (Muller and Cadwell, 1986) at the south ends of the Finger Lakes (Fig. 2). The Valley Heads appear to represent a relatively stable ice margin and the outpouring of large volumes of both local and exotic debris with overflow drainage to the south. Mullins and Hinchey (1989) suggested that much of the Valley Heads debris may have been transported and deposited by subglacial meltwaters issuing from beneath the southern margin of the Laurentide ice sheet (Fig. 45).

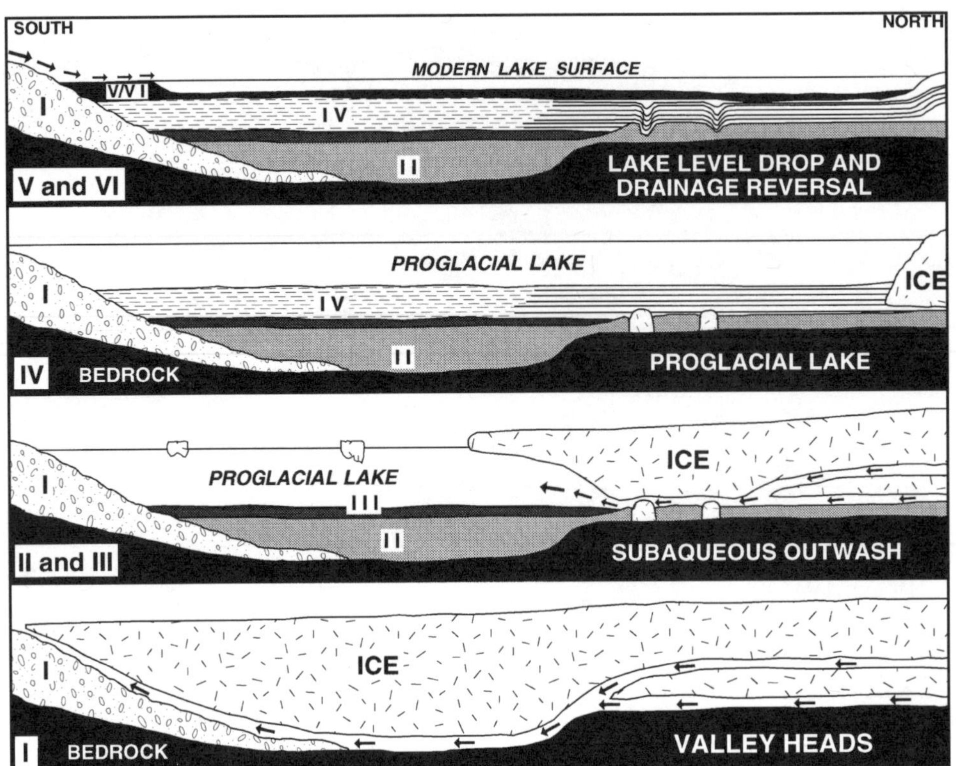

Figure 45. Schematic, longitudinal reconstruction of sediment infill history of the Finger Lakes. Panels from bottom to top represent successive stages of infill extending from Valley Heads deposition (sequence I, ~14.4 ka) to drainage reversal (sequence V, ~13.9 ka) resulting from an abrupt drop in proglacial lake levels.

Depositional sequence II beneath the Finger Lakes is a thick unit characterized by strong lateral changes in acoustic facies from north to south. It typically is thickest and finer grained to the south than it is to the north. Esker-like ridges have also been identified in association with sequence II, strongly suggesting an active role for subglacial meltwaters in the deposition of this sequence. The occurrence of dropstones in sequence II deposits at the Canandaigua drillsite further suggests the presence if icebergs shed from calving ice margins that likely fronted along deep, proglacial lakes. Following retreat of the ice margin from its Valley Heads position, sequence II appears to represent a subaqueous outwash phase of rapid outpouring of relatively fine-grained sediment into the Finger Lake basins (Fig. 45).

Depositional sequence III beneath the Finger Lakes is unique in that it is a highly transparent (reflection-free) acoustic unit indicating homogenous deposits. Although sequence III is the only unsampled Finger Lake depositional sequence, its transparency suggests that it consists of fine-grained sediments that were deposited rapidly. Sequence III typically is thickest in the central portions of the lakes and does not occur to the north. Sequence III sediments may represent lifting of the ice margin from its pinned sequence II position, and the very rapid outpouring of fine-grained sediment from beneath the ice sheet. However this interpretation must be considered speculative until sequence III sediments are sampled.

Depositional sequence IV is a distinctive acoustic unit in that it is characterized by high-frequency reflections that can be traced the entire length of the Finger Lake basins. The fact that sequence IV is found throughout the basins suggests that the lakes were uncovered by ice. Piston cores from sequence IV recovered fine-grained, rhythmites with occasional dropstones indicative of proglacial lake deposition. Although the lakes were uncovered during deposition of sequence IV, the ice margin was likely still close by, as indicated by the presence of dropstones. The ice margin was likely located along a series of till moraines that extend across the north end of the Finger Lakes (Figs. 42 and 46). At this time the ice formed dams at the north end of the lakes, resulting in elevated proglacial lake levels relative to today's lake levels.

Depositional sequence V represents the first evidence for significant sediment input to the Finger Lake basins from lateral and southerly sources. Prior to sequence V deposition, sediments entered the basins from the ice sheet to the north. However, once the ice sheet retreated northward, opening lower elevations to the north, proglacial lake levels in the region appear to have dropped precipitously (Hand and Muller, 1972; Hand, 1978), creating a drainage reversal. At this time sedi-

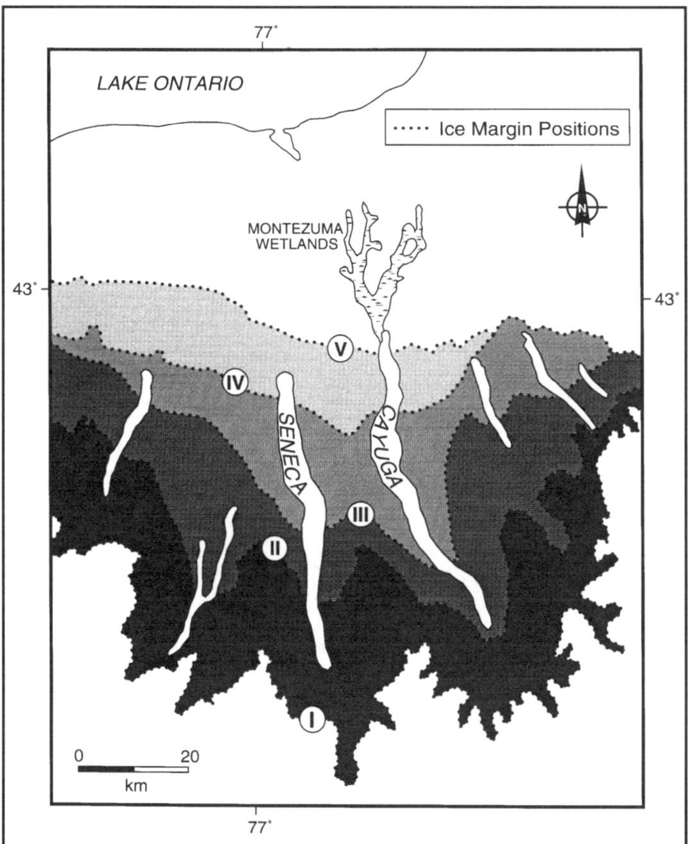

Figure 46. Map reconstruction of recessional ice margin positions in the eastern Finger Lakes region extending from ~14.4 ka (Valley Heads) to ~13.9 ka. Roman numeral designations refer to ice margin positions for depositional sequences in Seneca Lake. Note that lakes did not uncover simultaneously; Keuka was the first to be uncovered by ice and Cayuga the last. (Map based on surficial geologic map of Muller and Cadwell, 1986 [see Fig. 2], as well as data presented in this chapter. Constructed in consultation with E. H. Muller.)

ments began entering the Finger Lake basins from the south and sides as they do today. The well-known glens, gorges, and waterfalls of the Finger Lakes region (von Engeln, 1961) likely formed at this time as lateral streams adjusted to dramatically lowered local base levels. Drill results from both Canandaigua and Cayuga Valleys reveal that this event (lake-level drop) is recorded as a thick layer of sand and gravel, 10 to 15 m deep, which caps glaciolacustrine deposits beneath the valleys south of the lakes. A radiocarbon date of an in situ peat layer directly overlying these gravels indicates that lake levels dropped just prior to 13.6 ± 0.2 ka (Wellner et al., this volume). As lateral streams downcut their channels in response to this abrupt drop in lake level, alluvial fans spread out across the valleys south of the lakes and numerous deltas ("points") prograded into the Finger Lake basins from their margins.

Depositional sequence VI represent "modern" deposition in the Finger Lakes. These are mostly fine-grained, relatively organic-rich sediments that enter the basins from local watersheds. Sequence VI deposition probably began once hydrologic equilibrium had been achieved following the abrupt drop of proglacial lake levels.

Relationship to Heinrich event H-1

The thick sediment fill (up to 270 m) beneath the Finger Lakes was deposited rapidly over a short interval of time following deep erosion (to as much as 306 m below sea level) of the Finger Lake basins. Regionally constrained radiocarbon dates indicate that the Valley Heads (sequence I) were deposited sometime between 14.8 and 13.6 ka, most likely ~14.4 ka (Muller and Calkin, 1993). Radiocarbon dates from just above the sequence V-VI contact in Canandaigua Valley of 13.6 ± 0.2 ka (Wellner and Dwyer, this volume) and just below this contact in Seneca Lake of 13.9 ± 0.7 ka (King et al, 1983) restrict the age of the Finger Lakes sediment-fill to a millennium between ~14.4 and 13.5 ka. And, a recently obtained radiocarbon date from the V-VI contact in Seneca Lake of 13,865 ± 100 yr further restricts deposition of sequences I-V to a 500-yr-interval between 14.4 and 13.9 ka.

Deposition of the Valley Heads Moraine and thick sediment-fill beneath the Finger Lakes 14.4 to 13.9 ka is coeval with Heinrich event H-1 (Broecker, 1994). In 1988 Heinrich reported a series of coarse-grained, ice-rafted debris layers found in piston cores from the northeast Atlantic Ocean that he linked to the release of icebergs from the Laurentide ice sheet. Broecker et al. (1992) radiocarbon-dated Heinrich layer H-1 at ~15 ka and argued that the Heinrich events were a consequence of surges along the eastern margin of the Laurentide ice sheet (Broecker et al., 1992).

Bond et al. (1992) documented that the short-lived, massive discharges of icebergs that occurred during each Heinrich event were also accompanied by dramatic decreases in sea-surface temperature, surface salinity, and planktonic foraminifera. They further constrained the age of H-1 to between 14,590 ± 230 and 13,490 ± 220 yr. Bond et al. (1992) also noted that events H-1 and H-2 (~21 ka) correlate with evidence for ice streaming in the Hudson Strait. Andrews and Tedesco (1992) also invoked ice streaming and general instability of the Laurentide ice sheet to explain Heinrich layers H-1 and H-2 in the Labrador Sea, which they found to be rich in Paleozoic carbonate fragments derived from eastern Canada. They dated H-1 at between 14,560 ± 105 and 13,185 ± 190 yr (Andrews and Tedesco, 1992) and subsequently documented that H-1 was associated with a major meltwater event (Andrews et al., 1994). Andrews and Tedesco (1992) argued that Heinrich layers H-1 and H-2 are evidence for very rapid oscillations (advances and retreats) of the Laurentide ice sheet, which may occur on time scales of only a few hundred years (Kaufman et al., 1993). Dowdeswell et al. (1995) have suggested that the most likely duration for Heinrich events is 250 to 1,250 yr. Correlation of the deposition of the Valley Heads and thick sediment-infill of the Finger Lakes with Heinrich event H-1 suggests that H-1 lasted less than 500 yr (14.4 to 13.9 ka).

The massive outpourings of icebergs from the Laurentide ice sheet needed to deposit the Heinrich layers must involve the rapid flow of ice, and apparently meltwater, to the margin of the ice sheet. However, the driving mechanism of these instabilities of the ice sheet remains uncertain. Although Oerlemans (1993) concluded that it was "unlikely" that Heinrich events are a direct response to climatic cooling, Bond et al. (1993) noted that Heinrich events are the culminations of "cooling cycles" followed by rapid transitions to warmer climate, perhaps associated with massive collapse of ice sheets. If Heinrich events are terminations of "cooling cycles," then they may have been driven by variations in North Atlantic thermohaline circulation (Bond et al., 1992), which strongly influences high northern latitude temperatures (Broecker et al., 1990; Broecker, 1994).

Alternate hypotheses for the origin of Heinrich events have focused on internal dynamics of ice sheets. MacAyeal (1993) suggested that the Laurentide ice sheet ". . . periodically disgorged icebergs in brief but violent episodes (Heinrich events) . . ." in response to "free (unforced) oscillations" of the ice sheet. He proposed a binge/purge model in which the ice sheet slowly grew while frozen to a rigid bed (binge phase), followed by a brief purge phase of ice streaming and discharge of icebergs once its base was lubricated via geothermal flux and basal ice melt (MacAyeal, 1993).

Hughes (1992) has also argued that ice streams (both marine and terrestrial) have life cycles that culminate in ice sheet collapse, resulting in iceberg outbursts (marine) and dust storms (terrestrial), which in turn drive abrupt climate change. Hughes (1992) has suggested that unstable ice streams were common along margins of the Laurentide ice sheet once it had advanced beyond crystalline bedrock of the Canadian Shield onto softer sedimentary rocks, particularly along river valleys and marine troughs. This concept of unstable ice sheet behavior over deformable beds has been further developed by Clark (1994). He has argued that rapid oscillations of the Laurentide ice sheet were not climatically controlled but rather were associated with inherent instabilities of the ice sheet during retreat over areas of deformable sediment. Clark (1994) has suggested that during unstable behavior ice margins may advance ". . . hundreds to thousands of meters per year over periods of hundreds of years" (p. 20). In the case of its southern margin, the Laurentide ice sheet advanced 300 to 800 km between about 15 and 14 ka, correlative with Heinrich event H-1 (Clark, 1994).

Most recently, Blanchon and Shaw (1995) have suggested that Laurentide ice sheet instability and collapse during H-1 was triggered by the storage and release of large volumes of glacial meltwaters, recorded as catastrophic sea-level rises in coral reef sequences. These authors argued that collapse of the ice sheet reduced its surface elevation enough to affect atmospheric circulation which in turn affected the ocean heat pump to high latitudes.

In the Finger Lakes, Heinrich event H-1 was a short-lived (less than 500 yr) phenomenon when large volumes of first coarse-grained, then fine-grained, sediments were disgorged along the southern margin of Laurentide ice sheet. The fact that much of the Valley Heads is water-laid drift, coupled with the presence of buried eskers beneath some of the Finger Lakes, suggests that much of this sediment may have been transported to the ice margin by subglacial meltwaters (Mullin and Hinchey, 1989) (Fig. 45). Release of large volumes of meltwater along the southern margin of the Laurentide ice sheet at this time is supported by oxygen isotope data from the Gulf of Mexico that suggest that a major discharge of meltwater began 14.5 ka and culminated ~13 ka (Leventer et al., 1982). Based on drumlin distributions, Shaw and Gilbert (1990) have also proposed a large-scale meltwater event for the Finger Lakes region, although they provide no constraints (other than late Wisconsin) on its timing. The release of large volumes of meltwater along the southern margin of the Laurentide ice sheet may have caused it to surge by decoupling the ice sheet from its bed and promoting large-scale instability (e.g., Iverson et al., 1995).

Evidence for rapid late Wisconsin ice flow into the Finger Lakes region is found in the well-developed drumlin field just north of the Finger Lakes in the Ontario Lowland. White (1985) interpreted this drumlin field as a product of rapid water-rich surges, whereas Ridky and Bindschadler (1990) viewed the drumlin field as a result of the flow of outlet glaciers from an ice dome in the eastern Lake Ontario basin. Regardless of which interpretation one favors, the drumlins north of the Finger Lakes record a phase of rapid ice flow over a soft, deformable bed (e.g., Boulton and Hindmarsh, 1987; Boyce and Eyles, 1991).

Laurentide ice sheet reconstructions for ~14 ka (Fig. 47) suggest the presence of ice domes over James Bay and in southeastern Quebec (Hughes, 1987). Ice from these two domes apparently converged and flowed into the eastern Lake Ontario basin and then out into the Finger Lakes region, terminating along the Valley Heads moraine (Figs. 46 and 47). Mullins and Hinchey (1989) viewed this as a terrestrially based ice stream, whereas Ridky and Bindschadler (1990) have argued that it was an outlet glacier. At the same time, ice was rapidly streaming through the Hudson Strait, supplying icebergs for the ice-rafted debris of Heinrich layer H-1 (Bond et al., 1992; Andrews and Tedesco, 1992). We suggest that the coeval rapid ice flow through the Hudson Strait, as well as into the Finger Lakes region, and the concomitant disgorging of sediment along both the eastern and southern margin of the ice sheet, was the result of a single unstable phase of the Laurentide ice sheet. At this time the southern margin of the Laurentide ice sheet was highly irregular, with large lobes penetrating southward into the midcontinent (Fig. 47) (Hughes, 1987; Clark, 1994). Glacial events similar to those that occurred in the Finger Lakes have recently been reported from the Puget Lowland of Washington, where the Cordilleran ice sheet reached its maximum extent ~14.5 ka (Booth, 1994).

The Finger Lakes provide a continental record of ice dynamics along the southern margin of the Laurentide ice sheet during Heinrich event H-1. This record reveals a rapid, erosive ice advance (surge) followed by, or contemporaneous with, the

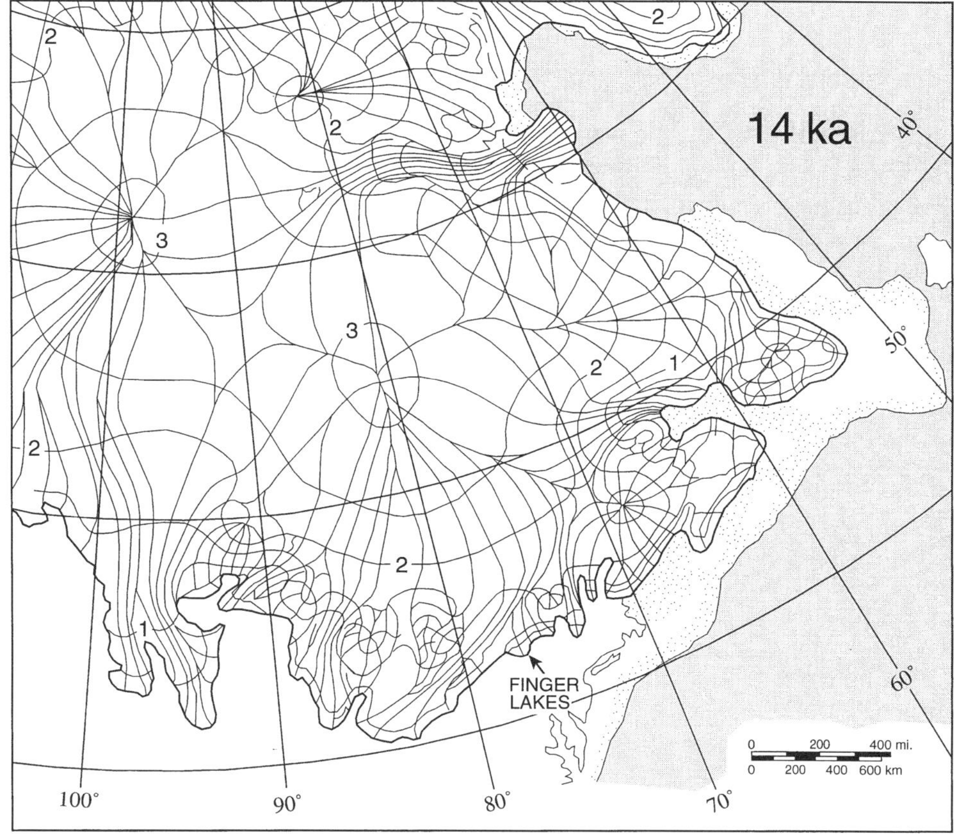

Figure 47. Reconstruction of Laurentide ice sheet ~14 ka (from Hughes, 1987). Note highly digitate southern margin of ice sheet as well as convergence of flow lines through the Hudson Straits (Heinrich event H-1) and into the Lake Ontario basin and the Finger Lakes region. Contours represent ice elevations in kilometers above sea level.

rapid outpouring of large volumes of sediment and meltwater. As such, it supports models that have called for large-scale instability and collapse of the Laurentide ice sheet ~14 ka. The fact that ice rapidly advanced and disgorged sediment along both the eastern (Heinrich layer H-1) and southern (Finger Lakes fill) margins of the Laurentide ice sheet supports suggestions (Clark, 1994) that this unstable behavior occurred on the scale of the entire ice sheet.

The driving mechanism of such unstable behavior of large ice sheets is not currently known but depends on whether different ice sheets respond independently (internal mechanism) or synchronously (external mechanism). Bond and Lotti (1995) have recently presented evidence for the synchronous discharge of icebergs from different North Atlantic sources during sub–Heinrich-scale events that they link with Dansgaard-Oeschger (cold-warm) cycles recorded in Greenland ice cores. Fronval et al. (1995) have also recently documented synchronous fluctuations of the Fennoscandian and Laurentide ice sheets during regional cooling and iceberg discharge events. And, in the western United States, alpine glaciers and ice caps appear to have been in phase with growth and collapse of the Laurentide ice sheet and North Atlantic Heinrich events (Clark and Bartlein, 1995). Collectively, these results argue for an atmospheric (climatic) forcing mechanism for Heinrich events of perhaps global extent (Broecker, 1994).

CONCLUSIONS

1. The Finger Lakes occupy deeply scoured, overdeepened glacial rock basins. Maximum scour occurs beneath Seneca Lake were bedrock has been eroded as much as 306 m (1,004 ft) *below* sea level. Whether this erosion has been the result of multiple glaciations cannot be determined by our study. However, late Wisconsin erosion did result in the removal of any preexisting materials and/or the downcutting of bedrock to depths as great as 306 m below sea level.

2. The Finger Lake basins have been infilled with a thick (up to 270 m) section of late Wisconsin sediment. These sediments were deposited as a series of six depositional sequences. Sequence I represents coarse-grained Valley Heads Moraine deposition ~14.4 ka. Sequences II and III represent the rapid outpouring of fine-grained outwash via subglacial meltwaters

along calving ice margins that fronted in deep, proglacial lakes. Sequence IV is a glaciolacustrine unit deposited in high-standing proglacial lakes once the ice margin had retreated to the northern ends of the lake basins. Sequence V marks a turning point in the sedimentary history of the Finger Lakes. Following ice retreat out of the basins, lake levels dropped precipitously, resulting in a regional drainage reversal and sediment input from the south by 13.9 ka. Glens, waterfalls, and deltas developed as lateral streams adjusted to an abrupt lowering of local base levels. Sequence VI records the focusing of relatively organic-rich postglacial sediments to the deep portions of the Finger Lakes.

3. The Valley Heads Moraines and sediment sequences I through V beneath the lakes were deposited rapidly during a short interval of time between 14.4 and 13.9 ka.

4. Late Wisconsin erosion and sediment-infill of the Finger Lakes was coeval with Heinrich event H-1 in the North Atlantic.

5. The Finger Lakes provide a continental record of a rapid erosive ice advance (surge?) followed by the disgorging of large volumes of meltwater and sediment along the southern margin of the Laurentide ice sheet.

6. These results support models of large-scale instability and rapid flow of ice within the Laurentide ice sheet ~14.5 to 14.0 ka.

ACKNOWLEDGMENTS

This research was supported by National Science Foundation Grants EAR–8607326 and EAR–8903870. We thank Ernie Muller for valuable discussion over the years. John Ladd assisted with wide-angle reflection experiments in the lakes, and Bob Sheridan provided onland multichannel seismic reflection equipment as well as advice during the drilling. We also thank John Petruccione for collection and processing of onland reflection data. Reviews by Paul Carlson (U.S. Geological Survey, Menlo Park) and Nick Eyles (University of Toronto) were of great value in revising this manuscript.

REFERENCES CITED

Andrews, J. T., and Tedesco, K., 1992, Detrital carbonate-rich sediments, northwestern Labrador Sea: Implications for ice sheet dynamics and iceberg rafting (Heinrich) events in the north Atlantic: Geology, v. 20, p. 1087–1090.

Andrews, J. T., Erlenkeuser, H., Tedesco, K., Aksu, A. E., and Jull, A. J. T., 1994, Late Quaternary (Stage 2 and 3) meltwater and Heinrich events, northwest Labrador Sea: Quaternary Research, v. 41, p. 26–34.

Bentley, C. R., 1987, Antarctic ice streams: A review: Journal of Geophysical Research, v. 92, p. 8843–8858.

Birge, E. A., and Juday, C., 1914, A limnological study of the Finger Lakes of New York: U.S. Bureau of Fisheries Bulletin, v. 32, p. 525–609.

Blanchon, P., and Shaw, J., 1995, Reef-drowning events during the last deglaciation: Evidence for catastrophic sea-level rise and ice-sheet collapse: Geology, v. 23, p. 4–8.

Bloom, A. L., 1984, Unanswered questions about Finger Lakes geomorphology: Cornell Quarterly, v. 19, p. 57–61.

Bloom, A. L., 1986, Geomorphology of the Cayuga Lake basin: New York State Geological Association Fieldguide 58, p. 261–279.

Bloomfield, J. A., 1978, Lakes of New York state. Vol. 1, Ecology of the Finger Lakes: New York, Academy Press, 499 p.

Bond, G. C., and Lotti, R., 1995, Iceberg discharges into the North Atlantic on millennial time scales during the last deglaciation: Science, v. 267, p. 1005–1010.

Bond., G., and 13 others, 1992, Evidence for massive discharges of icebergs into the North Atlantic Ocean during the last glacial period: Nature, v. 360, p. 245–249.

Bond, G., Broecker, W. S., Johnsen, S., McManus, J., Labeyrie, L., Jouzel, J., and Bonani, G., 1993, Correlations between climate records from North Atlantic sediments and Greenland ice: Nature, v. 365, p. 143–147.

Booth, D. B., 1994, Glaciofluvial infilling and scour of the Puget Lowland, Washington, during ice-sheet glaciation: Geology, v. 22, p. 695–698.

Boulton, G. S., and Hindmarsh, R. C. A., 1987, Sediment deformation beneath glaciers: Rheology and geological consequences: Journal of Geophysical Research, v. 92, p. 9059–9082.

Boyce, J. I., and Eyles, N., 1991, Drumlins carved by deforming till streams below the Laurentide ice sheet: Geology, v. 19, p. 787–790.

Broecker, W. S., 1994, Massive iceberg discharges as triggers for global climate change: Nature, v. 372, p. 421–424.

Broecker, W. S., Bond, G., and Klas, M., 1990, A salt oscillator in the glacial Atlantic? The concept: Paleoceanography, v. 5, p. 469–477.

Broecker, W. S., Bond, G., Klas, M., Clark, E., and McManus, J., 1992, Origin of the northern Atlantic's Heinrich events: Climate Dynamics, v. 6, p. 265–273.

Bryan, G. M., 1980, The hydrophone-pinger experiment: Journal of the Acoustical Society of America, v. 68, p. 1403–1408.

Clark, P. U., 1994, Unstable behavior of the Laurentide ice sheet over deforming sediment and its implications for climate change: Quaternary Research, v. 41, p. 19–25.

Clark, P. U., and Bartlein, P. J., 1995, Correlation of the late Pleistocene glaciation in the western United States with North Atlantic Heinrich events: Geology, v. 23, p. 483–486.

Clarke, G. K. C., 1987, Fast glacier flow: Ice streams, surging, and tidewater glaciers: Journal of Geophysical Research, v. 92, p. 8835–8841.

Coates, D. R., 1968, Finger Lakes, *in* Fairbridge, R. W., ed., Encyclopedia of geomorphology: New York, Reinhold Co., p. 351–357.

Coates, D. R., 1974, Reappraisal of the glaciated Appalachian Plateau, *in* Coates, D. R., ed., Glacial geomorphology: Binghamton, New York, State University of New York Publications, p. 205–243.

Davis, M. B., and Ford, M. S. J., 1982, Sediment focusing in Mirror Lake, New Hampshire: Limnology and Oceanography, v. 27, p. 137–150.

Dowdeswell, J. A., Maslin, M. A., Andrews, J. T., and McCave, I. N., 1995, Iceberg production, debris rafting, and the extent and thickness of Heinrich layers (H-1, H-2) in North Atlantic sediments: Geology, v. 23, p. 301–304.

Eyles, N., Mullins, H. T., and Hine, A. C., 1990, Thick and fast: Sedimentation in a Pleistocene fiord lake of British Columbia, Canada: Geology, v. 18, p. 1153–1157.

Eyles, N., Mullins, H. T., and Hine, A. C., 1991, The seismic stratigraphy of Okanagan Lake, British Columbia: A record of rapid deglaciation in a deep "fiord-lake" basin: Sedimentary Geology, v. 73, p. 13–41.

Fairchild, H. L., 1909, Glacier waters in central New York: State University of New York Education Department Bulletin 442, 66 p.

Fairchild, H. L., 1934a, Cayuga Valley lake history: Geological Society of America Bulletin, v. 45, p. 233–280.

Fairchild, H. L., 1934b, Senecy Valley physiographic and glacial history: Geological Society of America Bulletin, v. 45, p. 1073–1110.

Faltyn, N. E., 1957, Seismic exploration of the Tully overburden [M.S. thesis]: Syracuse, New York, Syracuse University, 86 p.

Finckh, P., Kelts, K., and Lambert, A., 1984, Seismic stratigraphy and bedrock forms in perialpine lakes: Geological Society of America Bulletin, v. 95, p. 1118–1128.

Fleisher, P. J., 1986, Dead-ice sinks and moats: Environments of stagnant ice deposition: Geology, v. 14, p. 39–42.

Fleisher, P. J., Mullins, H. T., and Yuretich, R. F., 1992, Subsurface stratigraphy of Otsego Lake, New York: Implications for deglaciation of the northern Appalachian Plateau: Northeastern Geology, v. 14, p. 203–217.

Fronval, T., Jansen, E., Bloemendal, J., and Johnsen, S., 1995, Oceanic evidence for coherent fluctuations in Fennoscandian and Laurentide ice sheets on millenium timescales: Nature, v. 374, p. 443–446.

Fullerton, D. S., 1986, Stratigraphy and correlation of glacial deposits from Indiana to New York and New Jersey: Quaternary Science Reviews, v. 5, p. 23–29.

Grousset, F. E., Labeyrie, L., Sinko, J. A., Cremer, M., Bond, G., Duprat, J., Cortijo, E., and Huon, S., 1993, Patterns of ice-rafted detritus in the glacial North Atlantic (40–55°N): Paleoceanography, v. 8, p. 175–192.

Hand, B. M., 1978, Syracuse meltwater channels: New York State Geological Association Fieldtrip Guidebook 50, p. 286–314.

Hand, B. M., and Muller, E. H., 1972, Syracuse channels: Evidence of a catastrophic flood: New York State Geological Association Fieldtrip Guidebook 44, p. I-1–12.

Heinrich, H., 1988, Origin and consequences of cyclic ice rafting in the northeast Atlantic Ocean during the past 130,000 years: Quaternary Research, v. 29, p. 142–152.

Hughes, T., 1987, Ice dynamics and deglaciation models when ice sheets collapsed, *in* Ruddiman, W. F., and Wright, H. E., eds., North America and adjacent oceans during the last deglaciation: Boulder, Colorado, Geological Society of America, Geology of North America, v. K-3, p. 183–220.

Hughes, T., 1992, Abrupt climate change related to unstable ice-sheet dynamics: Toward a new paradigm: Palaeogeography, Palaeoclimatology, Palaeoecology (Global and Planetary Change Section), v. 97, p. 203–234.

Iverson, N. R., Hanson, B., Hooke, R. LeB., and Jansson, P., 1995, Flow mechanism of glaciers on soft beds: Science, v. 267, p. 80–81.

Karrow, P. F., Warner, B. G., Miller, B. B., and McCoy, W. D., 1990, Reexamination of an interglacial section on the west shore of Cayuga Lake, New York: American Quaternary Association Abstracts with Programs, Waterloo, Ontario, p. 22.

Kaufman, D. S., Miller, G. H., Stravers, J. A., and Andrews, J. T., 1993, Abrupt early Holocene (9.9–9.6 ka) ice-stream advance at the moth of Hudson Strait, Arctic Canada: Geology, v. 1063–1066.

King, J. W., Banerjee, S. K., Marvin, J., and Lund, S., 1983, Use of small scale–amplitude paleomagnetic fluctuations for correlation and dating of continental climatic changes: Palaeogeography, Palaeoclimatology, Palaeoecology, v. 42, p. 167–183.

King, L. W., Rokoengen, K., Fader, G. B. J., and Gunleiksrud, T., 1991, Till-tongue stratigraphy: Geological Society of America Bulletin, v. 103, p. 637–659.

Krall, D. B., 1977, Late Wisconsin ice recession in east-central New York: Geological Society of America Bulletin, v. 88, p. 1697–1710.

Lehman, J. T., 1975, Reconstructing the rate of accumulation of lake sediment: The effect of sediment focusing: Quaternary Research, v. 5, p. 541–550.

Leventer, A., Williams, D. F., and Kennett, J. P., 1982, Dynamics of the Laurentide ice sheet during the last deglaciation: Evidence from the Gulf of Mexico: Earth and Planetary Science Letters, v. 59, p. 11–17.

Ludlam, S. D., 1967, Sedimentation in Cayuga Lake, New York: Limnology and Oceanography, v. 12, p. 618–632.

MacAyeal, D. R., 1993, Binge/purge oscillations of the Laurentide ice sheet as a cause of the North Atlantic's Heinrich events: Paleoceanography, v. 8, p. 775–784.

Maury, C. J., 1908, An interglacial fauna found in Cayuga Valley and its relation to the Pleistocene of Toronto: Journal of Geology, v. 6, p. 565–567.

Monnett, V. E., 1924, The Finger Lakes of central New York: American Journal of Science, v. 8, p. 33–53.

Muller, E. H., and Cadwell, D. H., 1986, Surficial geologic map of New York–Finger Lakes sheet: Albany, New York State Museum, Geological Survey Map and Chart Series No. 40, scale 1:250,000, 1 sheet.

Muller, E. H., and Calkin, P. E., 1993, Timing of Pleistocene glacial events in New York state: Canadian Journal of Earth Sciences, v. 30, p. 1829–1845.

Muller, E. H., and Prest, V. K., 1985, Glacial lakes in the Ontario Basin, *in* Karrow, P. F., and Calkin, P. E., Quaternary evolution of the Great Lakes: Geological Association of Canada Special Paper 30, p. 213–229.

Mullins, H. T., and Hinchey, E. J., 1989, Erosion and infill of New York Finger Lakes: Implications for Laurentide ice sheet deglaciation: Geology, v. 17, p. 622–625.

Mullins, H. T., Eyles, N., and Hinchey, E. J., 1990, Seismic reflection investigation of Kalamalka Lake: A "fiord lake" on the Interior Plateau of British Columbia: Canadian Journal of Earth Sciences, v. 27, p. 1225–1235.

Mullins, H. T., Eyles, N., and Hinchey, E. J., 1991a, High-resolution seismic stratigraphy of Lake McDonald, Glacier National Park, Montana, U.S.A.: Arctic and Alpine Research, v. 23, p. 311–319.

Mullins, H. T., Wellner, R. W., Petruccione, J. L., Hinchey, E. J., and Wanzer, S., 1991b, Subsurface geology of the Finger Lakes region, *in* Ebert, J. R., ed., New York State Geological Association, 63rd Annual Meeting Field Trip Guidebook: Oneonta, State University of New York, p. 1–54.

Oerlemans, J., 1993, Evaluating the role of climate cooling in iceberg production and Heinrich events: Nature: v. 364, p. 783–786.

Proctor, B. L., 1978, Chemical investigation of sediment cores from four minor Finger Lakes of new York [Ph.D. thesis]: Buffalo, State University of New York, 227 p.

Ridky, R. W., and Bindschadler, R. A., 1990, Reconstruction and dynamics of the late Wisconsin "Ontario" ice dome in the Finger Lakes region, New York: Geological Society of America Bulletin, v. 102, p. 1055–1064.

Schaffner, W. R., and Oglesby, R. T., 1978, Limnology of eight Finger Lakes: Hemlock, Canadice, Honeoye, Keuka, Seneca, Owasco, Skaneateles and Otisco, *in* Bloomfield, J. A., ed., Lakes of New York state. Vol. 1, Ecology of the Finger Lakes: New York, Academic Press, p. 313–470.

Schubel, J. R., and Schiemer, E. W., 1973, The cause of the acoustically impenetrable, or turbid, character of Chesapeake Bay sediments: Marine Geophysical Researches, v. 2, p. 61–71.

Shaw, J., and Gilbert, R., 1990, Evidence for large scale subglacial meltwater flood events in southern Ontario and northern New York State: Geology, v. 18, p. 1169–1172.

Stephens, D. B., 1986, Seismic stratigraphic analysis of glacial and post-glacial sediments in northern Seneca Lake, New York [M.S. thesis]: Syracuse, New York, Syracuse University, 145 p.

Tarr, R. S., 1894, Lake Cayuga: A rock basin: Geological Society of America Bulletin, v. 5, p. 339–356.

Tarr, R. S., 1904, Artesian well section at Ithaca, N.Y.: Journal of Geology, v. 12, p. 69–82.

von Engeln, O. D., 1929, Interglacial deposit in central New York: Geological Society of America Bulletin, v. 40, p. 469–479.

von Engeln, O. D., 1961, The Finger Lakes region: Its origin and nature: Ithaca, New York, Cornell University Press, 130 p.

White, W. A., 1985, Drumlins carved by rapid water-rich surges: Northeastern Geology, v. 7, p. 161–166.

Woodrow, D. L., 1978, Surface and near-surface sediments in the northern part of Seneca Lake, N.Y.: New York State Geological Association Fieldtrip Guidebook 50, p. 250–255.

Woodrow, D. L., Blackburn, T. R., and Monahan, E. C., 1969, Geological, chemical and physical attributes of sediments in Seneca Lake, New York, *in* Proceedings, Twelfth Conference on Great Lakes Research, p. 380–396.

Wright, H. W., Kutzbach, J. W., Webb, T., III, Ruddiman, W. F., Street-Perrott, F. A., and Bartlein, P. J., eds., 1993, Global climates since the last glacial maximum: Minneapolis Press, 569 p.

Manuscript Accepted by the Society January 16, 1996

Printed in U.S.A.

Geological Society of America
Special Paper 311
1996

Correlation of drillcore and geophysical results from Canandaigua Lake Valley, New York: Evidence for rapid late-glacial sediment infill

Robert W. Wellner* and John L. Petruccione*
Department of Earth Sciences, Heroy Geology Laboratory, Syracuse University, Syracuse, New York 13244
Robert E. Sheridan
Department of Geological Sciences, Wright-Reiman Geological Laboratory, Rutgers University, New Brunswick, New Jersey 08903

ABSTRACT

Correlation of drillcore (119 and 13 m) results, downhole geophysical logs, and onland seismic reflection profiles to an extensive (~120 km) high-resolution lake seismic reflection data set have been used to delineate the depositional environments and processes that have occurred within the Canandaigua Lake basin during the last deglaciation.

Several distinct depositional units have been recognized beneath the dry lake valley: (1) a basal coarse-grained sand and gravel (unit 1) at a depth of ~142 to 116 m that would not support open hole drilling; (2) a massive clay devoid of dropstones (unit 2a) at 116 to 94 m and a coarse-grained clayey to sandy gravel (unit 2b) at 94 to 83 m; (3) a rhythmically bedded silt and sand with dropstones (unit 4) at 83 to 27 m; (4) a washed sand and gravel unit with artesian water (unit 5) at 27 to 12 m; and (5) cyclically interbedded peat, marl, and clay (unit 6) at 12 m to the surface.

The basal sand and gravel of depositional unit 1 have been correlated, via drillcore, gas- and water-well, and geophysical data, with the Valley Heads Moraine and associated drift that crops out ~6 km south of Canandaigua Lake. Regional correlations, coupled with radiocarbon dates, indicate that the Valley Heads Moraine was deposited between 14,800 and 14,400 ^{14}C B.P. An onlapping relationship of the overlying sediments onto the Valley Heads drift implies that the entire fill sequence preserved within the basin was deposited during a single Late Wisconsin/Holocene event. A radiocarbon age of 13,650 ± 210 ^{14}C B.P. from the in situ peat of depositional unit 6, which is in contact with the alluvial sand and gravel of depositional unit 5, further confines the deposition of units 2 through 5 to the millennium between ~14,500 and 13,500 ^{14}C B.P. The radiocarbon ages and stratigraphic relationships that bracket the depositional age of units 2 through 5 (104 m of fill) are equivalent to those reported for Heinrich Event H-1 in the North Atlantic Ocean. Therefore it is proposed that the thick (up to 200+ m), rapidly deposited fill sequence beneath Canandaigua Lake, and most likely all the Finger Lakes, is a continental equivalent to this short-lived marine event, and records an unstable phase of Laurentide Ice Sheet retreat.

*Present addresses: Wellner, Exxon Production Research Company, P.O. Box 2189, Houston, Texas 77252-2189; Petruccione, Geophysics Division, Berkshire Environmental Associates, Inc., 409 Penn Avenue, Sinking Spring, Pennsylvania 19608-1109.

Wellner, R. W., Petruccione, J. L., and Sheridan, R. E., 1996, Correlation of drillcore and geophysical results from Canandaigua Lake Valley, New York: Evidence for rapid late-glacial sediment infill, *in* Mullins, H. T., and Eyles, N., eds., Subsurface Geologic Investigations of New York Finger Lakes: Implications for Late Quaternary Deglaciation and Environmental Change: Boulder, Colorado, Geological Society of America Special Paper 311.

INTRODUCTION

The Finger Lakes region of central New York State consists of 11 elongate, glacially scoured lake basins (Fig. 1). The lakes vary considerably in maximum length (5 to 61 km), maximum water depth (9 to 186 m), and lake elevation (118 to 334 m) (Schaffner and Oglesby, 1978). The history of highstand proglacial lake stages within individual basins, as well as the glacial sediments, depositional processes, and geomorphic landforms between basins has been extensively studied ever since 1868 when Agassiz spoke of the "glacial heritage" of the Finger Lakes regions (Coates, 1968; reviewed by Mullins et al., 1989). However, beyond limited borehole data from the valleys south of Cayuga and Seneca Lakes (Tarr, 1904) and the publication of a single interpreted line drawing of a seismic reflection profile from Seneca Lake (Woodrow et al., 1969), little was known about the geology beneath the lakes themselves (Mullins et al., 1991).

In an effort to resolve questions about the evolution of the Finger Lakes, Mullins and Hinchey (1989) conducted a comprehensive (~1,300 km) seismic reflection survey of the lakes. They discovered that the Finger Lakes basins have been eroded to as much as 304 m below sea level and subsequently infilled with a single, late Wisconsin through Holocene sequence that is up to 275 m thick.

In order to evaluate the Finger Lakes for potential drillcor-

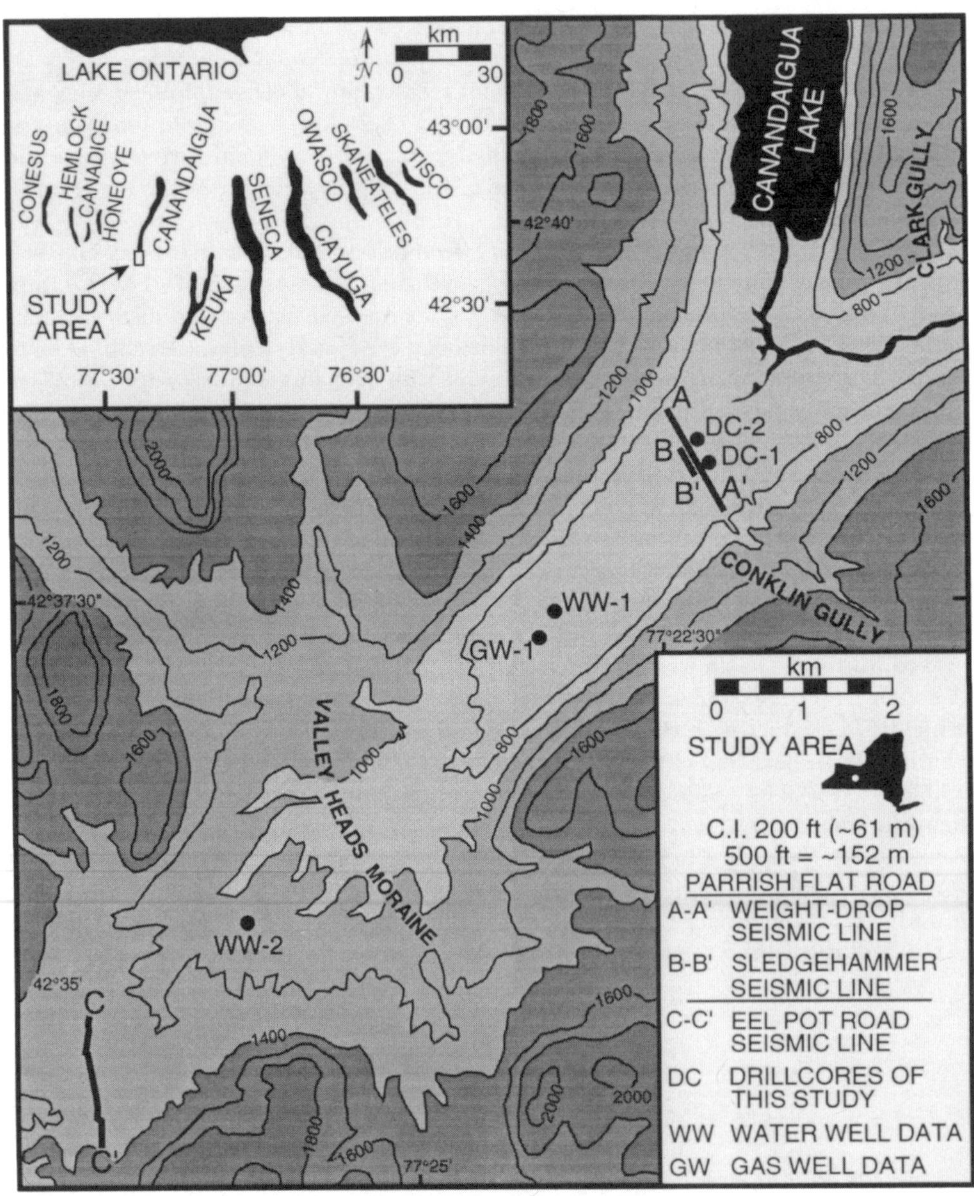

Figure 1. General index and study area location map. Note position of drillcores, gas wells, water wells, lakes, and onland seismic reflection tracklines. Contours given in feet.

ing, an extensive onland, multichannel seismic reflection survey was conducted in the valleys south of Skaneateles, Cayuga, Seneca, Canandaigua, and Hemlock Lakes (Fig. 1). On the basis of these data, two drillcores (119 and 13 m) were collected from the valley 3 km south of Canandaigua Lake near the town of Naples (Fig. 1). The primary purpose of this subsurface sampling was to (1) verify the geophysical data via correlation with subsurface samples, (2) reconstruct the first-order depositional history of Canandaigua Lake and the Finger Lakes region in general, on the basis of detailed core analysis, and (3) evaluate the thick sediment sequence for a record of deglacial processes and paleoenvironmental change at Canandaigua Lake.

SETTING

Canandaigua Lake occupies an axially overdeepened closed basin along the northwestern margin of the glaciated Appalachian Plateau. The lake is ~25 km long, 1 to 2.5 km wide, and up to 84 m deep (Eaton and Kardos, 1978) (Fig. 1).

To the south of Canandaigua Lake, a relatively flat valley extends ~8 km before the Valley Heads Moraine rises over 200 m above the valley floor (Fig. 1). The Valley Heads Moraine is confined to the Finger Lakes basins and consists of a massive complex of kame, kame moraine, and glaciofluvial deposits that has been dated on the basis of morphostratigraphy (Fullerton, 1986) and by radiocarbon (Muller and Calkin, 1993) between 14,400 and 14,100 ^{14}C B.P. In eastern New York State, the Valley Heads Moraine truncates the radiocarbon-dated Cassville-Cooperstown Moraine, which confines the maximum age for deposition to <14,800 ^{14}C B.P. (Krall, 1977). South of the Valley Heads Moraine are extensive outwash plains that are part of the south-flowing Susquehanna drainage system.

The uplands consist of exposed bedrock along the lake margins and relatively thin till, which becomes more aerially extensive and thicker to the north (Muller and Cadwell, 1986). Chevron-like recessional moraines also occur on the uplands. North of Canandaigua Lake are semiparallel, laterally discontinuous recessional moraines; highstand proglacial lake deposits (Fairchild, 1895), and an extensive field of drumlins and meltwater channels that extend to Lake Ontario (Muller and Cadwell, 1986).

METHODS

Seismic reflection profiles from Canandaigua Lake (~120 km) were collected aboard the *R/V Alexander Winchell* using an EG&G Uniboom sound source and a single, eight-element hydrophone streamer. The Uniboom profiler was fired with power outputs of 500 and 1,000-J, and a dominant frequency of ~1 kHz. All data were recorded as analog records. East-west–oriented "crossing" profiles were spaced at ~1-km intervals and tied by a north-south–oriented axial profile. Navigation was accomplished by using compass headings between known points and a constant boat speed., For a detailed seismic trackline map the reader is referred to Mullins et al. (this volume).

Lake seismic time sections were converted to depth sections using pinger-hydrophone wide-angle reflection data from Skaneateles Lake (Bryan, 1980; Mullins and Hinchey, 1989). These data suggest that the sediment velocity structure beneath the Finger Lakes varies in a linear fashion from 1.5 to 2.1 km/sec (shallow to deep). Individual sequences beneath Canandaigua Lake were assigned velocities that decrease linearly from 2.0 km/sec for Sequence I (oldest) to 1.5 km/sec for Sequence VI (youngest). A water column velocity of 1.455 km/sec was assumed., Subsurface depths and sequence thicknesses were then digitized, plotted, and contoured.

Two onland, multichannel, six-fold seismic reflection lines were collected in the valley 3 km south of Canandaigua Lake along Parrish Flat Road (Fig. 1). The sound source and receivers for this 1,060-m seismic line (Fig. 2) consisted of a 227-kg lead-shot–filled bag, which was dropped from a height of 3 m, and 8-Hz geophones. The shot offset and geophone spacings were both 20 m. A second, shorter 192-m seismic line was collected using a sledgehammer as a sound source and 60-Hz geophones as receivers. An optimum shot offset distance of 21 m and a geophone spacing of 3 m were employed., A third six-fold multichannel seismic reflection line (Fig. 3) was collected 10 km south of Canandaigua Lake on top of the Valley Heads Moraine (Fig. 1) using a 20-m shot offset and geophone spacing. A 12-channel Bison field seismograph was used to record the seismic data. The field data have been corrected for statics, prestack-filtered and gained, common-depth point (CDP) stacked, and poststack-filtered.

Two drillcores (5 cm in diameter) were collected from the valley ~3 km south of Canandaigua Lake (Fig. 1) using a mud rotary and wire-line coring technique. Site 1 extended 119 m into the subsurface; site 2, located 168 m to the west, extended 13 m into the subsurface (Fig. 1). The recovery was highly variable and averaged ~40% at drillsite 1 and 81% at drillsite 2. The drillcores were split, described, photographed, and radiographed.

Natural gamma-ray and resistivity logs were collected at coresite 1, before refusal at ~110 m due to collapse of the drillhole. Resistivity data were collected using electrode spacings of 0.406 and 1.63 m. Resistivity data could not be collected at subsurface depths of 0 to 27 m, due to the presence of steel casing that was inserted to stop the flow of water from an artesian aquifer between 12 and 27 m.

A vertical seismic profile (VSP) (see Balch and Lee, 1984) at drillsite 1 (Fig. 1) was also collected in the deep drillhole using a shotgun sound source and hydrophones, spaced at 1-m intervals, as receivers (Fig. 4). Shot offsets of 10 and 20 m were employed., The 10-m offset yielded better data and is thus presented in this chapter. The return signals were recorded on a 12-channel digital seismograph. The field data have had a static shift to align first arrivals, trace-gained, and frequency–wave number filtered to separate upgoing and downgoing events (Nobes and Schneider, this volume).

Synthetic seismograms for drillcore 1 were created using the GEOSIM synthetic seismogram program by converting

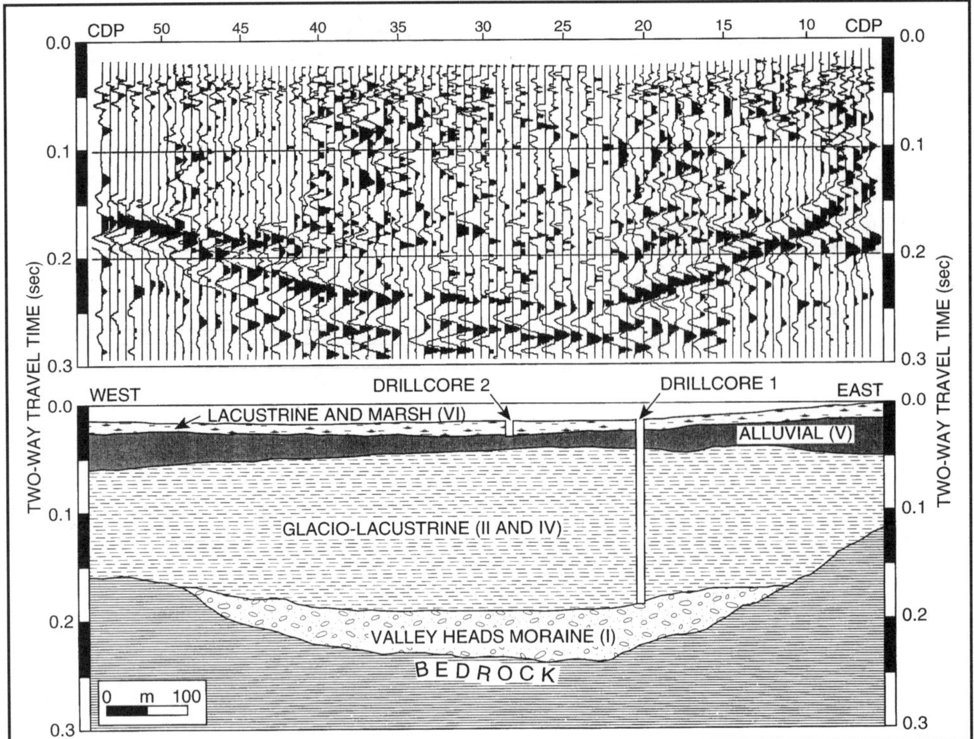

Figure 2. Weight-drop, multichannel seismic reflection profile A-A′ (top), and interpreted line drawing based on drillcore results (bottom) along Parrish Flat Road (Fig. 1). CDP = common-depth point.

raw resistivity logs and velocities of 1,700 m/sec for gravel and 1,500 m/sec for silt (e.g., Orsi and Dunn, 1991); the velocities for depositional units 6, 5, and bedrock were assumed to be ~1,000, 1,700, and 3.050 m/sec, respectively. These velocities agree with accepted values for the lithologies of the units involved (e.g., Orsi and Dunn, 1991), as well as the sediment velocities determined using the VSP data. Densities were computed from the input velocities by a Gardner function in the GEOSIM synthetic seismogram program. Densities and velocities were used to compute acoustic impedance contrasts and reflection coefficients at reflection times corresponding to depths at which lithologic contrasts occur. Time-linear logs of density, velocity, and reflection coefficient were plotted and compared with synthetic seismograms. The synthetic seismograms were produced by convolving a selected wave form or source pulse with the reflection coefficient time series.

RESULTS

Lithostratigraphy

A total of 75 core sections were recovered from the 119-m stratigraphic section drilled at core site 1 (Fig. 1). When core recovery was less than the cored interval, the top of the core was shifted to the beginning of the newly cored interval. Drillcore 2 (Fig. 1) had a total of seven core sections for the 13-m drilled interval.

On the basis of visual analysis and radiography, the subsurface lithostratigraphy south of Canandaigua Lake has been divided into five distinct, first-order lithostratigraphic units (Fig. 4). Lithostratigraphic units have been numbered to correspond with the six seismic sequences (I through VI) observed beneath Canandaigua Lake. As seismic sequence III does not extend beneath the southern dry valley, there is no lithostratigraphic unit 3.

Lithostratigraphic unit 1: Basal gravel (142–116 m). Basal unit 1 consists of coarse-grained gravel in a gray sandy-clay matrix (Fig. 5). Boulders larger in diameter than the drill bit were encountered in this unit, and as a result, drilling was terminated. The lower boundary of this sequence was determined via correlation of drill-core 1 to weight-drop seismic reflection data (Fig. 2).

Lithostratigraphic unit 2: Late-glacial sediments (116–83 m).

Subunit 2a (94–116 m). This subunit consists primarily of thick, massive gray clay with thin interbeds of silt and fine sand. This unit is devoid of dropstones and/or dropclots (Fig. 5). The thickness and grain size of the sand interbeds tend to increase near the base of this unit.

Subunit 2b (83–94 m). This subunit consists primarily of gray clayey-gravel and sandy-gravel, with thin interbeds of

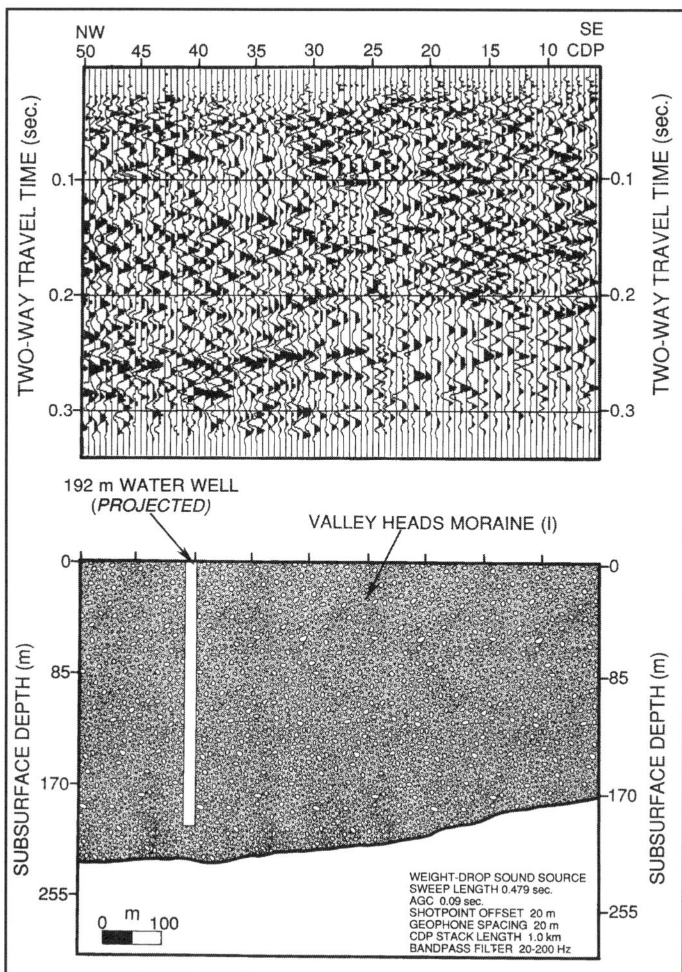

Figure 3. Weight-drop, multichannel seismic reflection profile C-C′ (top), and interpreted line drawing along Eel Pot Road (Fig. 1) on top of Valley Heads Moraine. Subsurface depth scale based on a P-wave velocity of 1,700 m/sec and correlation to water well (Fig. 1). CDP = common-depth point; AGC = automatic grain control.

gray-brown clayey-sand and brown coarse-grained sand (Fig. 5). The majority of the gravel consists of well-rounded, locally derived clasts.

Lithostratigraphic unit 4: Late-glacial sediments (27–83 m). Lithostratigraphic unit 4 consists of rhythmically interbedded gray brown to tan silt and silty sand and massive pink clay. Several coarser grained sand and sandy gravel layers are also present and interbedded with the massive clay. The spacing between interbeds and the number of dropstones decreases up core. This unit has been separated into four subunits (4a through 4d) on the basis of textural and bedding changes, abundance of dropstones, and correlation to the geophysical logs.

Subunit 4a (83–64 m). Unit 4a is characterized by thick-bedded, brown-gray massive clay with thin interbeds of tan silty clay and gray-brown sand (Fig. 6). The massive clay typically has numerous fine- to coarse-gravel–sized dropstones.

Subunit 4b (64–59 m). Unit 4b is characterized by massive gray-brown clay with thin brown, fine-grained sand interbeds (Fig. 6). Dropstones are sparse in this unit. Although there was no core recovery between the subsurface depths of 58 to 60 m, the interior of the core barrel was covered with coarse-grained sand and clayey gravel. This observation correlates well with shifts in the geophysical logs, which similarly indicate the presence of coarse-grained sediments within this interval.

Subunit 4c (59–54 m). The lower portion of unit 4c, between subsurface depths of 58 to 57 m, was not recovered. However, samples from subsurface depths shallower than 57 m recovered massive sandy gray clay with gravel. Based on correlation with the geophysical logs, the sediment in the missing interval is interpreted as sandy gray clay and gravel. Between subsurface depths of 57 to 54 m there is massive gray-brown clay with numerous sand- and gravel-sized dropstones (Fig. 6). This unit gradually becomes more well bedded and coarser grained by a subsurface depth of 56 m.

Subunit 4d (54–27 m). In general, this unit consists of interbedded, massive red-brown clay, tan silt, tan fine-grained sand, and brown gravelly sand (Fig. 7). There is a gradual change from clay-dominated layering between depths of 54 and 44 m to coarse-grained sand-dominated layering at depths shallower than 44 m. Above a depth of 40 m, the sandy interbeds thin (<2 cm) and become more silty, and by 37 m the bedding has thinned (<1 cm) and consists only of silt and clay. Bedding above a subsurface depth of 30 m has been disturbed by coring but appears to consist of thick, coarse-grained sand and thin, pink, silty clay interbeds. Dropstones are rare to absent throughout this interval.

Lithostratigraphic unit 5: Alluvial fan sediments (12–27 m). This unit consists of a coarsening-upward sequence of clayey-sand to washed gravel (Fig. 7). This interval contains artesian water with ~1.5 m of standing hydraulic head. Core recovery in this interval was reduced due to the presence of this artesian water and the coarse-grained nature of the sediments.

Lithostratigraphic unit 6: Post-glacial sediments (0–12 m). This unit consists of marl with freshwater gastropods; organic-rich clay with freshwater gastropods; peat; unfossiliferous, organic-poor clay; and an upper clayey soil horizon (Fig. 7). The peat at the base of unit 6 has been radiocarbon-dated at 13,650 ± 210 B.P. (University of Texas, Austin, sample 7253). Wellner and Dwyer (this volume) provide a detailed description of these sediments.

Geophysical logs

Because the geophysical logs extended to a subsurface depth of ~110 m, lithostratigraphic unit 1, the basal sand and gravel that terminated drillcoring, was not recorded. Between subsurface depths of 110 to 94 m, the sediments are characterized by high gamma-ray activity and low resistivity values (Fig. 4), which correlate to the interbedded massive brown-gray clay and sand of lithostratigraphic unit 2a (Fig. 4). Between subsurface depths of 94 and 81 m, the sediments are

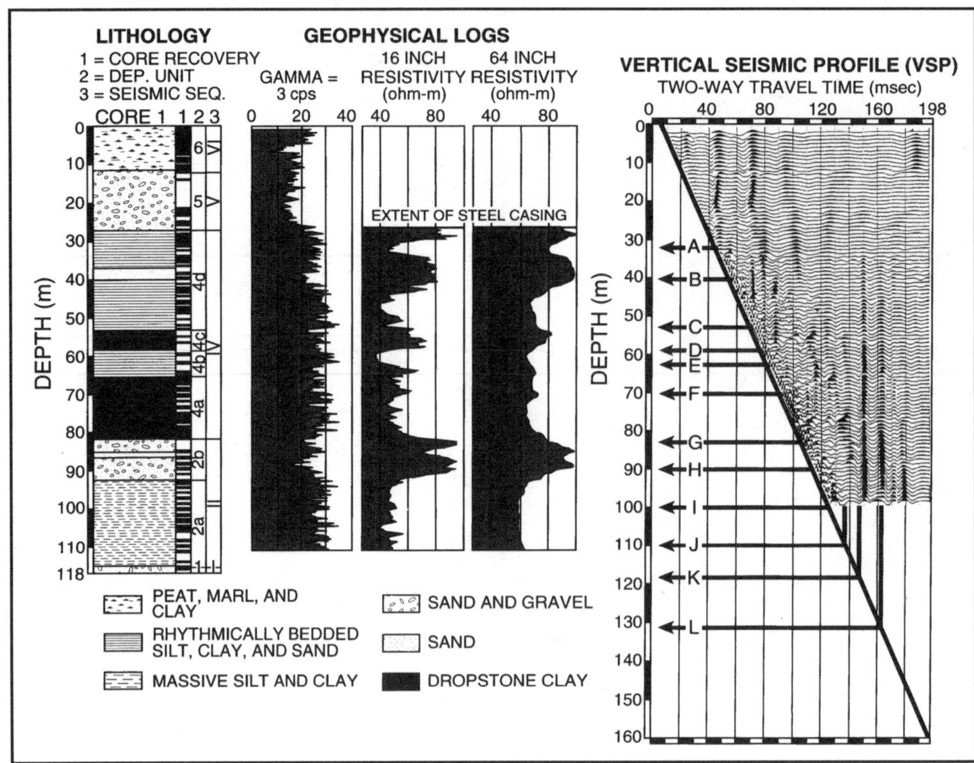

Figure 4. Correlation of first-order stratigraphy recovered beneath the lake valley 3 km south of Canandaigua Lake (Fig. 1) (left) with downhole geophysical data (center) and the vertical seismic profile (right). Note that refraction and acceleration of the down-going wavelet by the steel casing in drillhole 1 has occurred and has been corrected for by adding 10 msec of two-way travel time to each reflection event.

characterized by high resistivity and a somewhat lower gamma-ray activity relative to the underlying unit, which correlates to the clayey sand and gravel of lithostratigraphic unit 2b (Fig. 4). Between subsurface depths of 81 and 27 m, the sediments consist of widely varying resistivity values and gamma-ray activities that correlate to the rhythmically bedded proglacial lake sediments of lithostratigraphic unit 4 (Fig. 4). The abrupt signal shift at the depth of 27 m correlates to the alluvial sand and gravel of lithostratigraphic unit 5 (Fig. 4). Log unit 6, between the subsurface depths of 12 and 0 m, correlates to the postglacial lacustrine sediments of lithostratigraphic unit 6 (Fig. 4). This lithostratigraphic unit exhibits little change in gamma-ray activity from the underlying log unit, but has been well defined by coring. For a detailed description of the geophysical logs and their sedimentologic implications, the reader is referred to Nobes and Schneider (this volume).

Vertical seismic profile

A total of 14 seismic reflectors (A through N) were recorded by the vertical seismic profile (VSP) experiment (Fig. 4). A 10-msec correction factor has been applied to the travel time of each reflector, in order to correct for a casing pipe refraction that apparently accelerated the direct (D) wave; this in turn shifted events to shallower reflection times. Reflector N at 0.17 sec of travel time projects to a subsurface depth of 142 m (Fig. 4) and is interpreted, based on correlation with the multichannel seismic reflection data (Fig. 2), as bedrock. A second high-amplitude reflector above bedrock at 0.145 sec of travel time (reflector K, Fig. 4) correlates with the top of the basal gravel unit recovered in drillcore 1 at a subsurface depth of 118 m (Fig. 4) and to a high-amplitude reflector observed on onland, multichannel seismic reflection sections (Figs. 2 and 10). VSP reflectors H and G (Fig. 4) correlate with the contact between lithostratigraphic units 2a/2b and 2b/4a, respectively. VSP reflectors F-B (Fig. 4) correlate to coarser grained layers within the generally finer grained lithostratigraphic unit 4. Using the VSP technique it is possible to track the propagation of the output seismic signal as it penetrates the subsurface and observe reflections as they are generated along surfaces of impedance. Depths to reflecting interfaces are known from the drillcore record and travel times can be determined directly from the VSP, thus yielding an estimate for sediment velocities between reflecting surfaces (see Fig. 11). On the basis of the VSP data, interval velocities for the sediment-fill above bedrock ranged from <1,500 to 1,680 m/sec (Fig. 8) and have an average interval velocity of 1,560 m/sec.

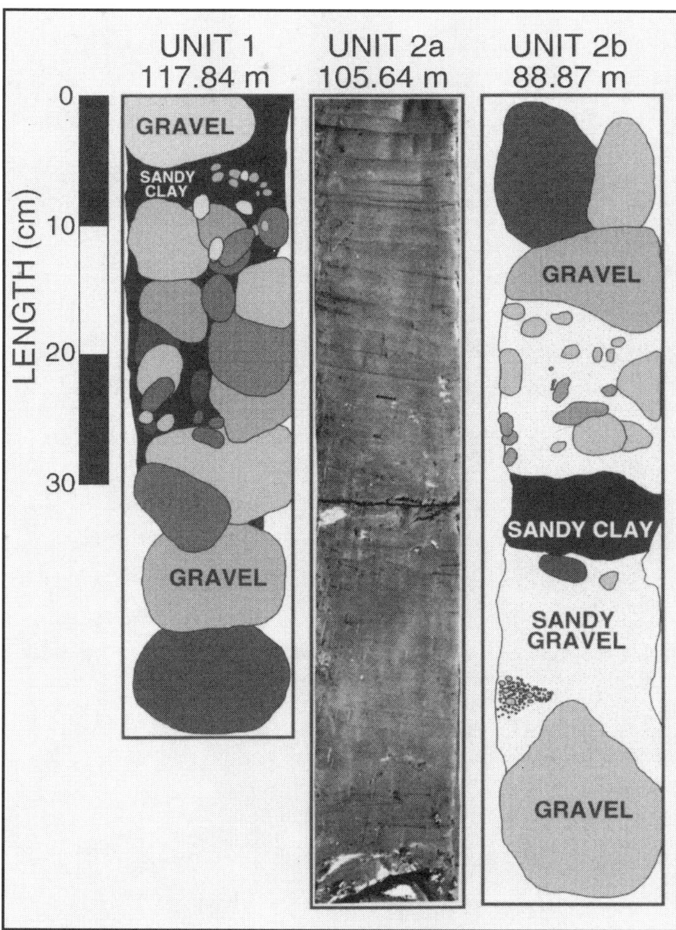

Figure 5. Core photographs typical of depositional units 1, 2a, and 2b core sediments referred to in text. Subsurface depths given at top of each core photograph.

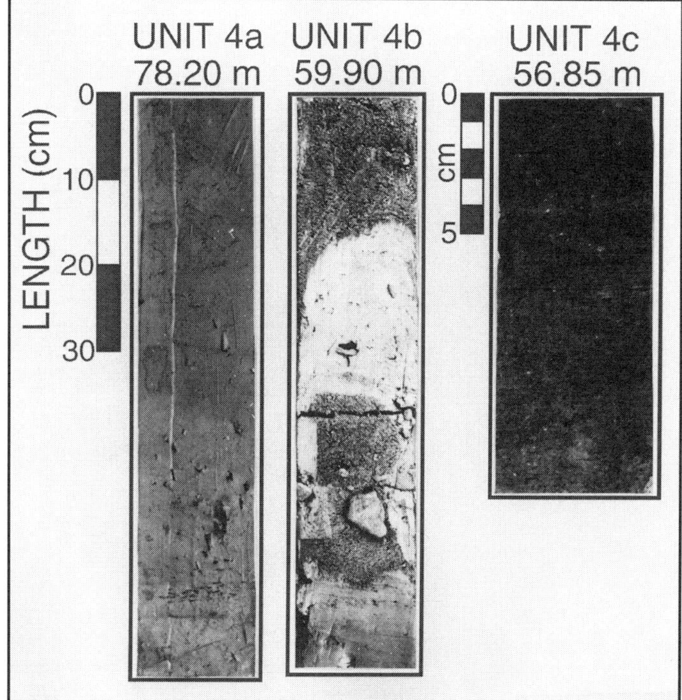

Figure 6. Core photographs typical of depositional units 4a, 4b, and 4c core sediments referred to in text. Subsurface depths given at top of each core photograph.

Lake seismic reflection data

Seismic sequence and facies analysis has identified six depositional sequences beneath Canandaigua Lake. The oldest sequence (I) consists of hummocky chaotic reflections resting above bedrock beneath the southern portion of Canandaigua Lake. Beneath Skaneateles Lake, sequence I can be traced directly to outcrops of Valley Heads drift (see Mullins et al., this volume), the upper portion of which has been radiocarbon dated at ~14,400 B.P. (Muller and Calkin, 1993).

Sequence II is characterized by chaotic, high-amplitude reflections in the north and an acoustically transparent facies in the south (Fig. 9). This sequence is interpreted as poorly sorted coarse-grained sediments in the north, which grade into massive, fine-grained deposits to the south. Sequence III is characterized by a transparent facies (Fig. 9) and is interpreted as rapidly deposited, homogeneous fine-grained sediments. Sequence IV consists of laterally continuous, parallel high- to low-amplitude reflections and is interpreted as rhythmically bedded, fine-grained sediments (Fig. 9). Sequence V is interpreted as distal, fine- (north) to proximal, coarse-grained (south) deltaic deposits. Sequence VI is characterized by low-amplitude, laterally continuous reflections and exhibits evidence of sediment focusing toward the central portion of the lake basin (Fig. 9). This sequence is interpreted as postglacial sediments. For an extensive description of the seismic stratigraphy beneath Canandaigua Lake, the reader is referred to the chapter by Mullins et al. (this volume).

Figure 10 is a 1,000-J axial profile of the southern portion of Canandaigua Lake (Fig. 1) illustrating the reflection characteristics and stratigraphic relationships among sequences II, IV, V, and VI. Sequence III is confined to the northcentral portion of the Canandaigua Lake basin (Fig. 9) and is not present on this seismic section. Note the apparent truncation of the seismic stratigraphy along the delta at the southern margin of the lake (Fig. 10). In all likelihood these seismic stratigraphic units continue beneath the modern delta and adjacent lake valley (Fig. 1).

Onland seismic reflection data

The multichannel, weight-drop seismic reflection profile collected along Parrish Flat Road (Fig. 1) indicates that top of bedrock has a maximum two-way travel time of 0.22 sec (Fig. 2). Assuming an average interval velocity of 1,560 m/sec, there is a maximum ~175 m of sediment above bedrock. Bedrock beneath drillsite 1 occurs at a subsurface depth of ~142 m. This profile also suggests the presence of four first-

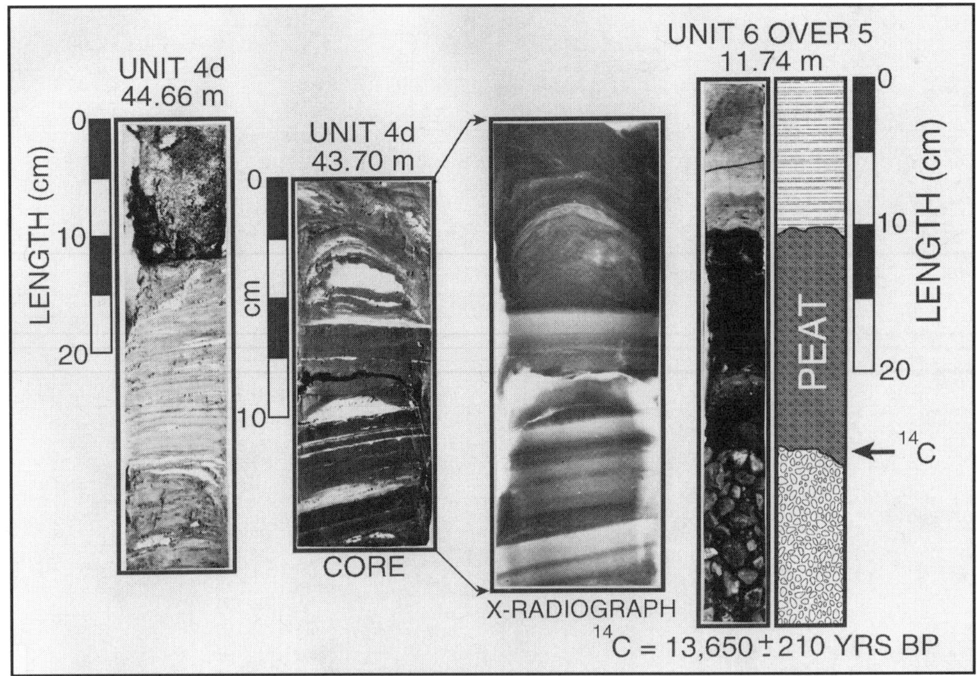

Figure 7. Core photographs and radiograph typical of depositional unit 4d sediments. Core photograph of the contact between the alluvial sand and gravel of depositional unit 5 and the peat of unit 6 is also shown. Note radiocarbon date from this contact. Subsurface depths given at top of each core photograph.

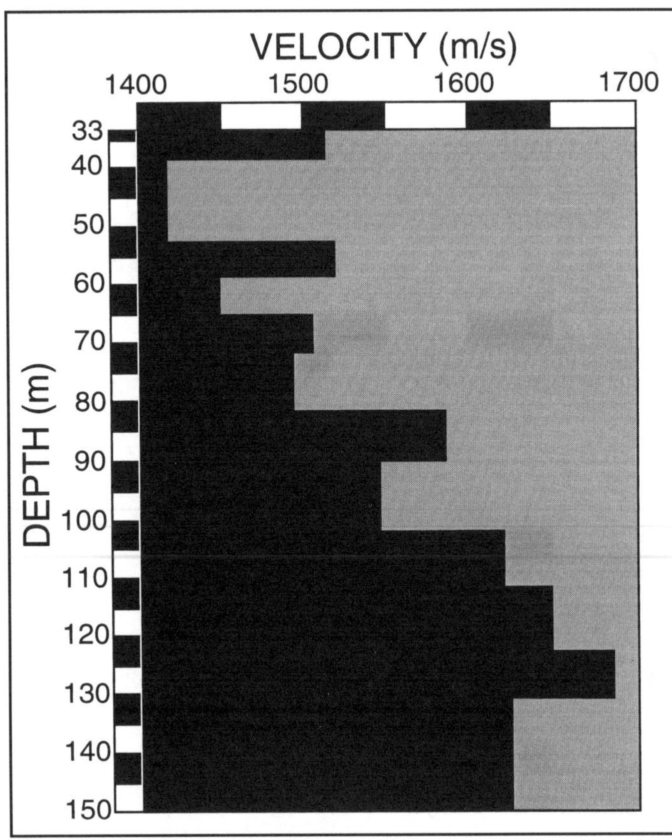

Figure 8. Velocity-depth profile for drillcore 1 sediments based on the vertical seismic reflection data.

order seismic units: a thin (~12 m), upper transparent unit underlain by a thin (~14 m), high-frequency unit that thins toward the center of the basin; a thick (~92 m) unit characterized by numerous, spurious high-amplitude reflections; and a basal unit beneath an irregular, high-amplitude, onlapping reflector (Fig. 2).

The high-resolution sledgehammer seismic reflection profile reveals the presence of as many as 13 laterally continuous reflectors, the majority of which can be correlated with contacts between depositional units (Fig. 11) and prominent reflectors on the lower resolution, weight-drop seismic data (Fig. 2). The correlation of the onland seismic data with the drillcores allows a three-dimensional stratigraphic framework to be developed for the fill beneath the dry lake valley. In general, the first-order depositional units observed at drillsite 1 are laterally continuous across the basin.

The multichannel seismic reflection profile collected along the Valley Heads Moraine (Figs. 1 and 3) indicates that bedrock is as much as 0.26 sec of two-way traveltime beneath the surface. The sediment fill above bedrock is characterized by chaotic, high-amplitude reflections, which is typical of poorly sorted and poorly stratified sediments. A water well drilled 7.8 km north of this seismic line (Fig. 1) penetrated 192 m of clay, fine-grained sand, and clayey gravel, and did not reach bedrock (A. Rendall, New York State Department of Conservation, personal communication, 1992). Assuming that these sediments are primarily water-saturated sand, which typically have

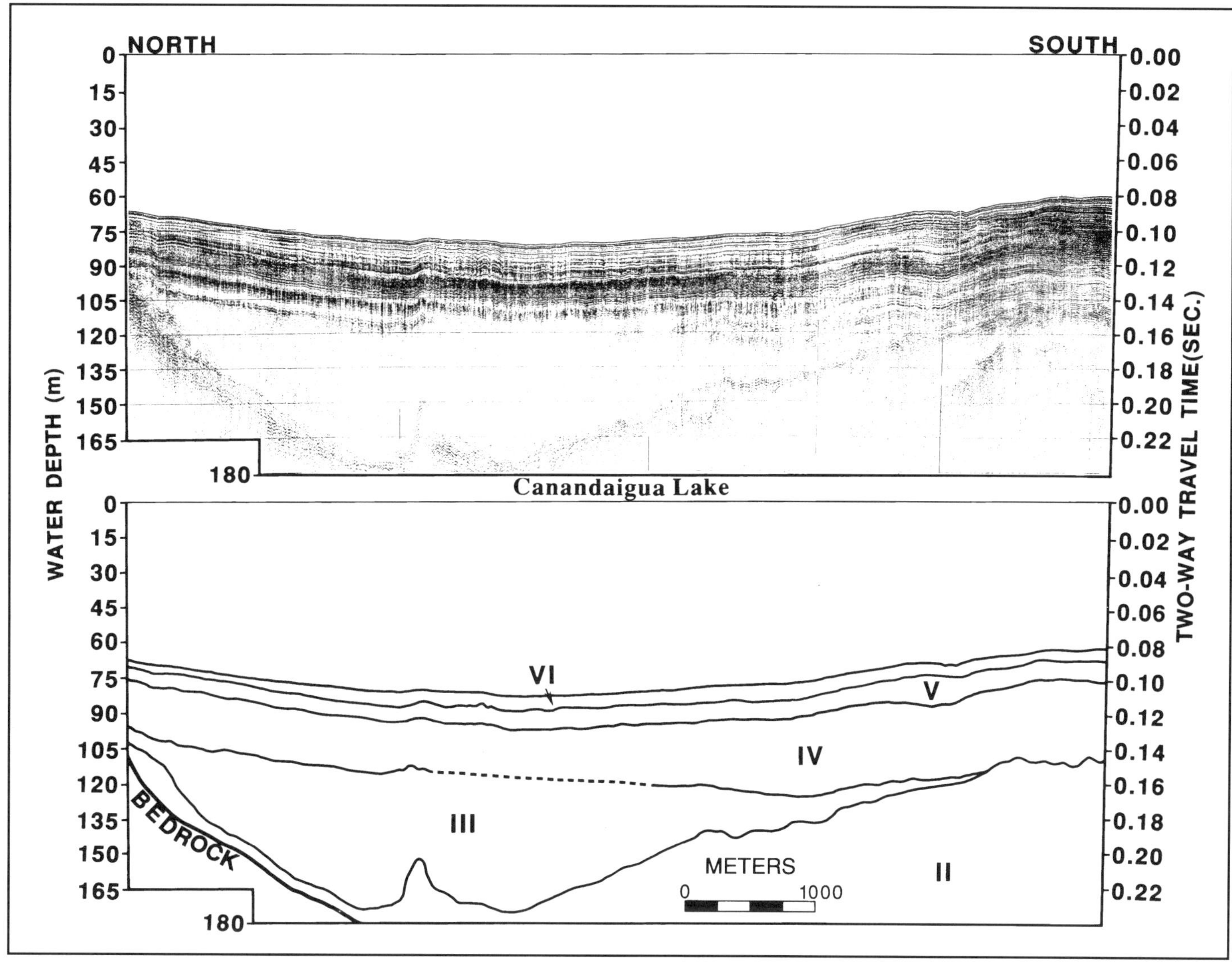

Figure 9. Uninterpreted (top) and line drawing (bottom) of axial (north-south), 1,000-J seismic reflection profile of the central portion of Canandaigua Lake. Vertical exaggeration is 22×. Water depths based on a P-wave velocity of 1,500 m/sec. Roman numerals refer to seismic sequences mentioned in text.

a velocity of 1,700 m/sec (Orsi and Dunn, 1991), there is ~220 m of sediment above bedrock.

Synthetic Seismograms

Three synthetic seismograms were produced to compare with the three data sets: (1) zero-phase Ricker wavelet at 45 Hz, to compare with the weight-drop seismic reflection data; (2) –90-phase causal (minimum phase) wavelet at 85 Hz, to compare with the sledgehammer seismic reflection data; and (3) a spike-wavelet at 999 Hz, to compare with the 1,000-Hz Uniboom seismic reflection data from Canandaigua Lake. In general, the synthetic seismograms, derived from core and log data, together with the onland and lake seismic data sets, exhibit similar reflection patterns and vertical facies transitions (Fig. 11). On the basis of these observations, the stratigraphic framework observed beneath the lake has been extended into the southern lake valley and allows for the comparison of core results to lake seismic data (Fig. 10).

DISCUSSION

Correlation of drillcore and VSP results with onland, multichannel seismic reflection data indicates that bedrock is ~142 m beneath the surface at drillsite 1 (Fig. 1) and reaches a maximum depth of ~175 m near the central portion of the basin (Fig. 2). Although bedrock was not reached by our drillcoring, a gas well located ~2.8 km south of drillsite 1, at the base of the Valley Heads Moraine, did encounter bedrock at a subsurface depth of 143 m (well record, Donohue, Anstey and Morrill, Inc., Boston, 1983) (Fig. 1). Borehole descriptions from this

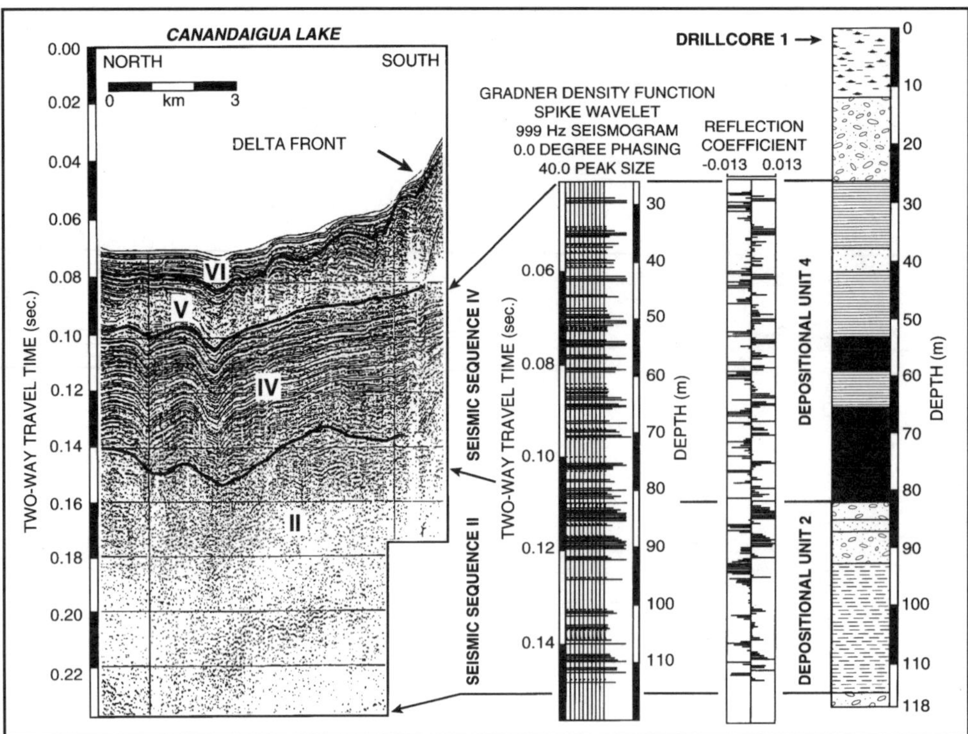

Figure 10. Correlation of 1,000-J seismic reflection profile from the southernmost end of Canandaigua Lake (left) to 999-Hz synthetic seismogram (center), which is based on downhole resistivity data and the first-order lithostratigraphy observed in drillcore 1 (right). Note depth scale is the same for stratigraphic column and synthetic seismogram section. See Figure 4 for stratigraphic legend.

gas well indicate that the sediments above bedrock are composed largely of fine-grained sand and gravel. An artesian aquifer was encountered above bedrock that caused water to rise within 5 m of the surface (137 m of hydraulic head). A water well, located ~400 m north of the gas well (Fig. 1), penetrated 18.6 m of interbedded coarse-grained sand and fine-grained to boulder-sized gravel. The sediments encountered in these wells are very similar to those reported from a deep (192 m) water well located on top of the Valley Heads Moraine (Fig. 1) (A. Rendall, New York State Department of Conservation, personal communication, 1992).

On the basis of core and seismic reflection data, Valley Heads drift descends beneath the surface ~6 km south of Canandaigua Lake (Fig. 1), and continues to the north where it correlates to the basal sand and gravel of lithostratigraphic unit 1 in drillcore 1 and to sequence I on lake seismic reflection sections (Fig. 12). Lithostratigraphic units 2 through 6 overlay the Valley Heads drift, and thus are younger than ~14,400 ^{14}C B.P. (Muller and Calkin, 1993). A radiocarbon date from an in situ peat layer at the contact between depositional units 5 and 6 (Fig. 9) further confines the age of depositional units 2 through 5 to between ~14,400 and 13,650 ± 210 ^{14}C B.P. (University of Texas, Austin, sample 7253). Based on the radiocarbon dates, units 2 through 5 (~104 m) were deposited at a rate of ~138 m/1,000 yr.

Correlation of drillcore sediments, via synthetic seismograms, with the extensive lake seismic data set can be used to extend the stratigraphic framework beneath the Canandaigua Lake basin. On the high-frequency (999 Hz) synthetic seismogram, the rhythmically bedded sediments of depositional unit 4 are characterized by a series of high-frequency reflections (Fig. 10). The reflection character of depositional unit 4 is very similar to that of lake seismic sequence IV, which is characterized by high-frequency, low- to high-amplitude, laterally continuous reflections (Fig. 10). The massive sediments of lithostratigraphic unit 2 are characterized by a largely transparent facies on the high-frequency (999 Hz) synthetic seismogram (Fig. 10). This reflection character is similar to that observed for lake seismic sequence II, which consists of a reflection-free facies in the southern end of Canandaigua Lake (Fig. 10). Based on the similarity between the synthetic and lake seismic data (Fig. 10), depositional units 4 and 2 are correlated with lake seismic sequences IV and II, respectively.

Synthetic seismograms (85 and 45 Hz) have also been used to correlate and extend the first-order lithostratigraphy observed in drillcore 1 and on gamma-ray and resistivity logs with sledgehammer (Fig. 11) and weight-drop seismic reflection data. In general, reflections observed on the lower resolution, onland seismic data correspond with substantial sedimentologic changes (e.g., unit 2a) associated with contacts between first-order depositional units in drillcore 1.

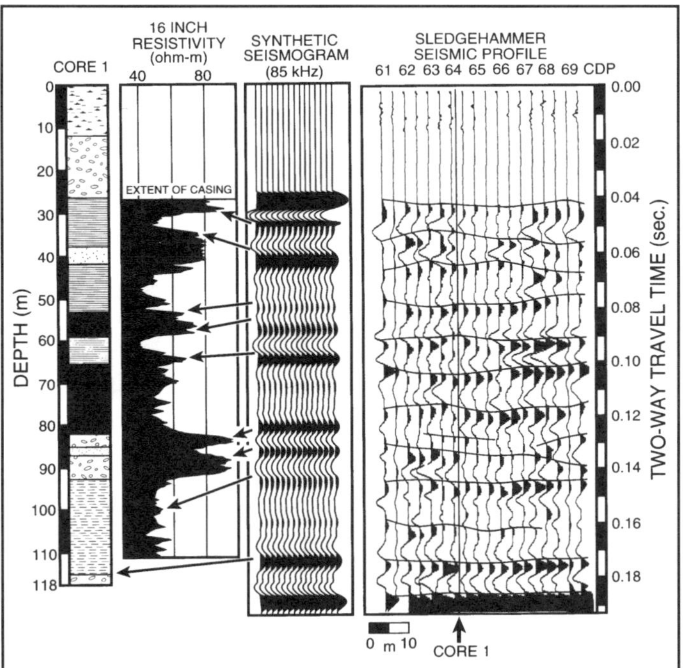

Figure 11. Correlation of first-order stratigraphy observed in drillcore 1 (see Fig. 4 for description of sediment types) to onland, multichannel sledgehammer seismic reflection data via the 16-in. resistivity log (left-center) and synthetic seismograms (right-center). Note depth scale is similar for stratigraphic column, resistivity log data, and synthetic seismogram. Onland, sledgehammer seismic section scale is in time.

The depositional record obtained in drillcore 1 appears, based on correlation with geophysical logs and onland seismic sections, to be complete and confirms the depositional model initially proposed by Mullins and Hinchey (1989) and expanded on by Mullins et al. (this volume), who argue for a single, late Wisconsin sedimentary sequence within each Finger Lake following the scour of the basins by rapidly flowing ice and subglacial meltwater, associated with the collapse of the Laurentide Ice Sheet. Models of the late Wisconsin Laurentide Ice Sheet (~14 ka) by Hughes (1987) similarly argue for the rapid outflow of ice from the Hudson Bay area, through the St. Lawrence Lowlands, and ultimately into the Finger Lakes region. Termination 1a in the marine oxygen isotope record also records the input of glacial meltwaters into the world's oceans at ~14,000 ^{14}C B.P. (Duplessy et al., 1981; Mix and Ruddiman, 1985; Berger et al., 1987) during the transition from glacial to interglacial climatic conditions.

Interestingly, the radiocarbon dates and stratigraphic relationships that bracket the deposition of sediments beneath Canandaigua Lake and the southern dry valley (14,400 to 13,650 ^{14}C B.P.) are similar to those reported by Bond et al. (1992) for Heinrich Event H-1 (14,590 to 13,490 ^{14}C B.P.). This is considered to be the result of massive and widespread input of ice-rafted debris from eastern Canada into the North Atlantic Ocean. We propose that the regionally extensive Valley Heads Moraine and thick, rapidly deposited sediments within the Canandaigua Lake basin provide a continental record of large-scale Laurentide Ice Sheet collapse that is equivalent to Heinrich Event H-1 in the Atlantic Ocean (Mullins and Wellner, 1993).

Following ice withdrawal from the Valley Heads Moraine, the basin was rapidly infilled with fine-grained glaciolacustrine sediments. Initially these sediments were deposited by subglacial meltwater as subaqueous outwash beneath Canandaigua Lake and the southern dry valley. The fine-grained, massive sediments of depositional unit 2 and the thick (>120 m) acoustically transparent sequence II on lake seismic sections are associated with this subaqueous deposit.

The morphology of this sequence and nature of depositional unit 2 sediments are consistent with observations from Alaskan fjords that exhibit subglacial meltwater discharge and rapid deposition from sediment plumes (Powell, 1981; Molnia, 1983; Gustavson and Boothroyd, 1987; Powell and Molnia, 1989; Cowan and Powell, 1991). Eventually, ice within the basin retreated to the northernmost portion of the basin. Water levels rose in response to this melting and retreat of the ice within the Canadaigua basin. This may be the early highstand proglacial Lake Naples of Fairchild (1895). From this position,

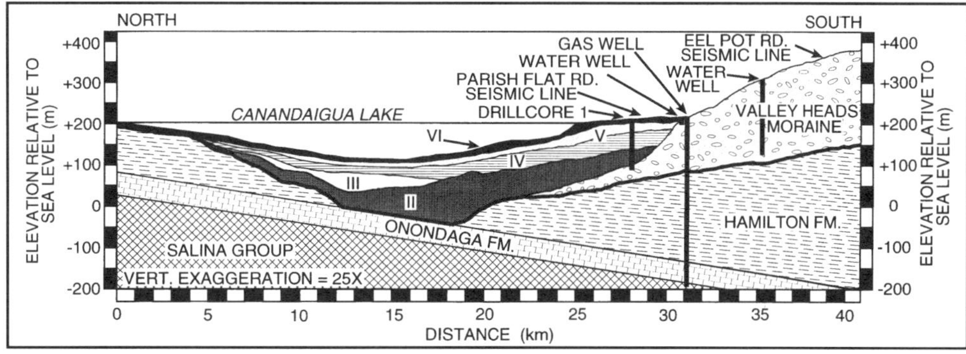

Figure 12. Schematic axial (north-south) profile of bedrock and stratigraphy of sediment-fill beneath Canandaigua Lake and the valley to the south. Note the location of drillcore 1, gas wells and water wells, and onland seismic reflection data.

lake seismic sequence III sediments are interpreted to have been rapidly deposited, via sediment gravity underflows, into a 50+ m-deep depression in the central portion of the basin. The rhythmically interbedded clay, silt, and sand of depositional unit 4 (subunits 4b and 4d) beneath the lake valley and the high-frequency, laterally continuous facies of lake seismic sequence IV are interpreted to have been deposited by seasonally driven gravity underflows. The thicker bedded sediments with numerous dropstones that characterize subunits 4a and 4c appear to record the more rapid delivery and deposition of ice-rafted debris from a calving ice front associated with the retreat of ice from this basin. The rhythmically bedded sediments of depositional unit 4 and lake seismic sequence IV are thickest beneath the dry lake valley (>60 m). This suggests that the primary sediment source was inwash off the uplands and Valley Heads Moraine to the south, rather than the Laurentide Ice Sheet. This source area reversal appears to occur slightly earlier at Canandaigua Lake than in the other Finger Lakes (Mullins and Hinchey, 1989). Continued ice retreat to the north eventually opened outlets for highstand proglacial lake waters to escape to the east, which rapidly lowered the water levels in the Canandaigua Lake basin. This also resulted in fluvial upland incision (see Conklin and Clark Gullies, Fig. 1), alluvial fan development, and rapid deltaic build-out. This interval of time is represented by depositional unit 5 in the dry lake valley, which consists of a coarsening-upward sequence of sand and gravel with artesian water, and by lake seismic sequence V beneath the lake. The incision of glens and gullies in the Canandaigua Lake basin, as well as in the other Finger Lakes, appears to have occurred extremely rapidly, over a very brief period prior to deposition of unit 6 at ~13,600 ^{14}C B.P. Deposition of unit 6 occurred in response to glacioisostatic rebound of the depressed northern outlet of this basin. This in turn resulted in reduced outflow of water from Canandaigua Lake, increased lake level, and reflooding of the recently exposed southern end of Canandaigua Lake (Tarr, 1904). This initial flooding and continued lake-level rise occurred after 13,650 ± 210 ^{14}C B.P. (University of Texas, Austin, sample 7253) and is represented by the cyclically interbedded peat, marl, and clay of depositional unit 6 beneath the dry lake valley and by sequence VI in the lake.

The stratigraphy beneath the Canandaigua Lake Valley is not unique. Drillcore results reported by Tarr (1904) for the valley south of Cayuga Lake near Ithaca, as well as available information from the Onondaga Trough (see Mullins et al., 1991), a dry Finger Lake–like valley ~12 km east of Otisco Lake (Fig.1), are very similar to that at Canandaigua Valley. This likely implies that similar depositional processes occurred within each Finger Lakes basin during the late Wisconsin collapse of the Laurentide Ice Sheet (Mullins et al., this volume).

SUMMARY AND CONCLUSIONS

1. On the basis of physical stratigraphy, geophysical logs, and onland seismic sections, five distinct depositional units have been recognized beneath the valley: (1) a basal coarse-grained sand and gravel (unit 1) at a depth of ~142 to 116 m, which would not support open hole drilling; (2) a massive clay unit devoid of dropstones at a depth of 116 to 94 m (unit 2a) and a coarse-grained clayey to sandy gravel (unit 2b) from 94 to 83 m; (3) a rhythmically bedded silt and sand with dropstones (unit 4) at a depth of 83 to 27 m; (4) a coarsening-upward washed sand and gravel with artesian water (unit 5) at a depth of 26 to 12 m ; and (5) cyclically interbedded peat, marl, and clay (unit 6) from a depth of 12 m to the surface.

2. On the basis of drillcore data, water- and gas-well data, and onland and lake seismic reflection data, the drift associated with the Valley Heads Moraine is interpreted to descend beneath the surface 6 km south of Canandaigua Lake, and continues to the north where it correlates with the basal depositional unit 1 beneath the dry valley and seismic sequence I beneath Canandaigua Lake. This indicates that the fill within the Canandaigua basin was deposited during a single, late Wisconsin/Holocene event since ~14,400 ^{14}C B.P. (Muller and Calkin, 1993). A radiocarbon date of 13,650 ± 210 ^{14}C B.P. (University of Texas, Austin, sample 7253) for a peat at the contact between depositional units 5 and 6 further confines the age of deposition of depositional units 2 through 5. On the basis of the radiocarbon ages and the stratigraphic relationships observed in the Canandaigua basin, we propose that the regionally extensive Valley Heads Moraine and thick, rapidly deposited sediments within the Canandaigua Lake and other Finger Lakes basins, is the onland equivalent of Heinrich Event H-1 observed in the North Atlantic Ocean.

3. Synthetic seismograms have been used to correlate lithostratigraphic units 2 and 4 observed in the drillcore to lake seismic sequences II and IV, respectively. Stratigraphic relationships indicate that depositional units 5 and 6 correlate with lake seismic sequences V and VI, respectively.

4. The similarity between the first-order drillcore results from the valley south of Canandaigua Lake and those previously reported from the valley south of Cayuga Lake and the Onondaga Trough (e.g., Tully Valley), coupled with the similarity between seismic reflection data sets from the lakes themselves, suggests that the first-order lithostratigraphy and depositional processes within the individual lake basins were uniform. Furthermore, the general depositional model documented for Canandaigua Lake may be similar for all the Finger Lakes basins.

ACKNOWLEDGMENTS

This research was supported by National Science Foundation Grant EAR–8703870 to H. T. Mullins. We thank David Nobes and George Schneider for collecting and processing the geophysical log data for Canandaigua Drillcore 1, and Parker Calkin and Peter Knuepfer for their comments on earlier drafts of this manuscript.

REFERENCES CITED

Balch, A. H., and Lee, M. W., 1984, Vertical seismic profiling: Technique, applications and case histories: Boston, International Human Resources Development Corporation, 488 p.

Berger, W. H., Killingley, J. S., and Vincent, E., 1987, Time scale of the Wisconsin/Holocene transition: Oxygen isotope record in the western Equatorial Pacific: Quaternary Research, v. 28, p. 245–306.

Bond, G., and 13 others, 1992, Evidence for massive discharges of icebergs into the North Atlantic Ocean during the last glacial period: Nature, v. 360, p. 245–249.

Bryan, G. M., 1980, The hydrophone-pinger experiment: Journal of Acoustical Society of America, v. 68, p. 1403–1408.

Coates, D. R., 1968, Finger Lakes, *in* Fairbridge, R. W., ed., Encyclopedia of geomorphology: New York, Reinhold Books, p. 351–356.

Cowan, E. A., and Powell, R. D., 1990, Suspended sediment transport and deposition of cyclically interlaminated sediment in a temperate glacial fjord, Alaska, USA, *in* Dowdesville, J. A., and Scourse, J. D., eds., Glacial marine environments: Processes and sediments: Geological Society of London Special Publication 53, p. 75–89.

Duplessy, J. C., Delibrias, G., Turon, J. L., Fujol, C., and Dupart, J., 1981, Deglacial warming of the northeastern Atlantic Ocean: Correlation with the paleoclimatic evolution of the European continent: Palaeogeography, Palaeoclimatology, Palaeoecology, v. 35, p. 121–144.

Eaton, S. W., and Kardos, L. P., 1978, The limnology of Canandaigua Lake, *in* Bloomfield, J. A., ed., Lakes of New York state: New York, Academic Press, p. 226–307.

Fairchild, H. L., 1895, Glacial lakes of western New York: Geological Society of America Bulletin, v. 6, p. 353–374.

Fullerton, D. S., 1986, Stratigraphy and correlation of glacial deposits from Indiana to New York and New Jersey: Quaternary Science Reviews, v. 5, p. 23–29.

Gustavson, T. C., and Boothroyd, J. C., 1987, A depositional model for outwash, sediment sources and hydrologic characteristics, Malaspina Glacier, Alaska: A modern analog of the southeastern margin of the Laurentide Ice Sheet: Geological Society of America Bulletin, v. 99, p. 187–200.

Hughes, T., 1987, Ice dynamics and deglaciation models when ice sheets collapse, *in* Ruddiman, W. F., and Wright, H. E., Jr., eds., North America and adjacent oceans during the last deglaciation: Boulder, Colorado, Geological Society of America, Geology of North America, v. K-3, p. 183–220.

Krall, D. B., 1977, Late Wisconsinan ice recession in east-central New York, Geological Society of America Bulletin: v. 88, p. 1697–1710.

Mix, A. C., and Ruddiman, W. F., 1985, Structure and timing of the last deglaciation: Oxygen isotope evidence: Quaternary Science Reviews, v. 4, p. 59–108.

Molnia, B. F., 1983, Subarctic glacial-marine sedimentation: A model, *in* Molnia, B. F., ed., Glacial marine sedimentation: New York, Plenum Press, p. 54–114.

Muller, E. H., and Cadwell, D. H., 1986, Surficial geologic map of New York–Finger Lakes sheet: Albany, New York Museum, Geologic Survey Map and Chart Series 40, 1 sheet, scale 1:250,000.

Muller, E. H., and Calkin, P. E., 1993, Timing of Pleistocene glacial events in New York state, Canadian Journal of Earth Sciences, v. 30, p. 1829–1845.

Mullins, H. T., and Hinchey, E. J., 1989, Erosion and infill of New York Finger Lakes: Implications for Laurentide Ice sheet deglaciation: Geology, v. 17, p. 621–625.

Mullins, H. T., Hinchey, E. J., and Muller, E. H., 1989, Origin of New York Finger Lakes: A historical perspective on the ice erosion debate: Northeastern Geology, v. 11, p. 166–181.

Mullins, H. T., Wellner, R. W., Petruccione, J. L., Hinchey, E. J., and Wanzer, S., 1991, Subsurface geology of the Finger Lakes region, *in* Ebert, J. R., ed., New York State Geological Association, 63rd Annual Meeting Field Trip Guidebook: Oneonta, State University of New York, p. 1–54.

Orsi, T. H., and Dunn, D. A., 1991, Correlations between sound velocity and related properties of glacio-marine sediments, Barents Sea: Geo-Marine Letters, v. 11, p. 79–83.

Powell, R. D., 1981, A model for sedimentation by tidewater glaciers: Annals of Glaciology, v. 2, p. 124–134.

Powell, R. D., and Molnia, B. F., 1989, Glacial marine sedimentation processes, facies, and morphology of the south-southwest Alaska shelf and fjords: Marine Geology, v. 85, p. 359–390.

Schaffner, W. R., and Oglesby, R. T., 1978, Limnology of eight Finger Lakes: Hemlock, Canadice, Honeoye, Keuka, Seneca, Owasco, Skaneateles, and Otisco, *in* Bloomfield, J. A., ed., Lakes of New York State, v. 1, Ecology of the Finger Lakes: New York, Academic Press, p. 313–471.

Tarr, R. S., 1904, Artesian well sections at Ithaca, NY: Journal of Geology, v. 12, p. 69–82.

Woodrow, D. L., Blackburn, T. R., and Monahan, E. L., 1969, Geological, chemical and physical attributes of sediments in Seneca Lake, New York, *in* Proceedings, Great Lakes Research Conference 12th, p. 380–396.

MANUSCRIPT ACCEPTED BY THE SOCIETY JANUARY 16, 1996

Printed in U.S.A.

Geological Society of America
Special Paper 311
1996

Results of downhole geophysical measurements and vertical seismic profile from the Canandaigua borehole of New York State Finger Lakes

David C. Nobes
Department of Geological Sciences, University of Canterbury, Private Bag 4800, Christchurch, New Zealand
George W. Schneider*
Department of Earth Sciences, University of Waterloo, Waterloo, Ontario N2L 3G1, Canada

ABSTRACT

A set of downhole geophysical logs and a vertical seismic profile (VSP) were acquired from the dry valley south of Canandaigua Lake, New York, to complement and calibrate shallow seismic profiles, and to supplement the subsurface drill cores. The upper 27 m of the borehole was cased with steel to seal an artesian layer; the remaining 91 m was left open for logging. For logistic reasons, the downhole logs were restricted to natural gamma and resistivity measurements; no active nuclear sources were used. Natural gamma readings were taken from the surface to a depth of 112 m, whereas resistivity data were recorded only below the steel casing. The results of the geophysical logging suggest that the drilled Quaternary sediments are composed predominantly of interbedded silt and sand or sandy gravel layers. The gamma and resistivity logs are well correlated; a principal component analysis helps in the creation of a quantitative lithology log and in the construction of a revised lithologic stratigraphy that consists of 5 first-order and 11 second-order units. The electrical properties are markedly different for the two electrode spacings, indicative of significant interbedding of thin layers. A vertical seismic profile was collected with a suite of 11 hydrophones placed every meter down the hole; a "shotgun" acoustic source was located 10 and 20 m from the borehole, similar to source-receiver offsets used in roadside reflection surveys. Bottom hole conditions were such that the VSP extended only to 99 m. The VSP results have been analyzed to separate the downgoing and upgoing energy. The variations in the geophysical logs correspond well to VSP reflectors. Although the steel casing severely degrades the upgoing VSP energy near the surface, the VSP section can still be used for calibration of the deeper seismic reflective sequence. The VSP reflectors have been processed, "flattened," and positioned for their appropriate two-way travel times. Identified reflectors correlate with glaciolacustrine units and subunits, the underlying Valley Heads Moraine, and the top of bedrock.

*Present address: Golder Associates Ltd., 2180 Meadovale Boulevard, Mississauga, Ontario L5N 5S3 Canada.

Nobes, D. C., and Schneider, G. W., 1996, Results of downhole geophysical measurements and vertical seismic profile from the Canandaigua borehole of New York State Finger Lakes, *in* Mullins, H. T., and Eyles, N., eds., Subsurface Geologic Investigations of New York Finger Lakes: Implications for Late Quaternary Deglaciation and Environmental Change: Boulder, Colorado, Geological Society of America Special Paper 311.

INTRODUCTION

The Finger Lakes of central New York State are long, linear glacial lakes that were scoured out and subsequently filled with up to 275 m of glaciolacustrine sediments during the last deglaciation (Mullins and Hinchey, 1989). Seismic reflection profiling of both the lakes and adjacent onshore areas has been conducted in order to evaluate the Quaternary subsurface stratigraphy of the region (Mullins et al., 1991). During July 1990, drillcore was collected from a 118-m-deep borehole located in the dry valley 3 km south of Canandaigua Lake (Fig. 1), in order to correlate subsurface stratigraphy with seismic reflection profiles and to collect samples for paleoenvironmental analysis (see Wellner et al., this volume). In support of the drilling operation, we collected a set of downhole geophysical logs and a vertical seismic profile. Our objectives were to provide a direct correlation with, and calibration of, seismic reflection profiles, and to collect continuous downhole measurements of the electrical and natural gamma response of the sediments to complement the incomplete core recovery and thus provide a more complete subsurface stratigraphy. The results and interpretation of the downhole geophysical measurements are reported here.

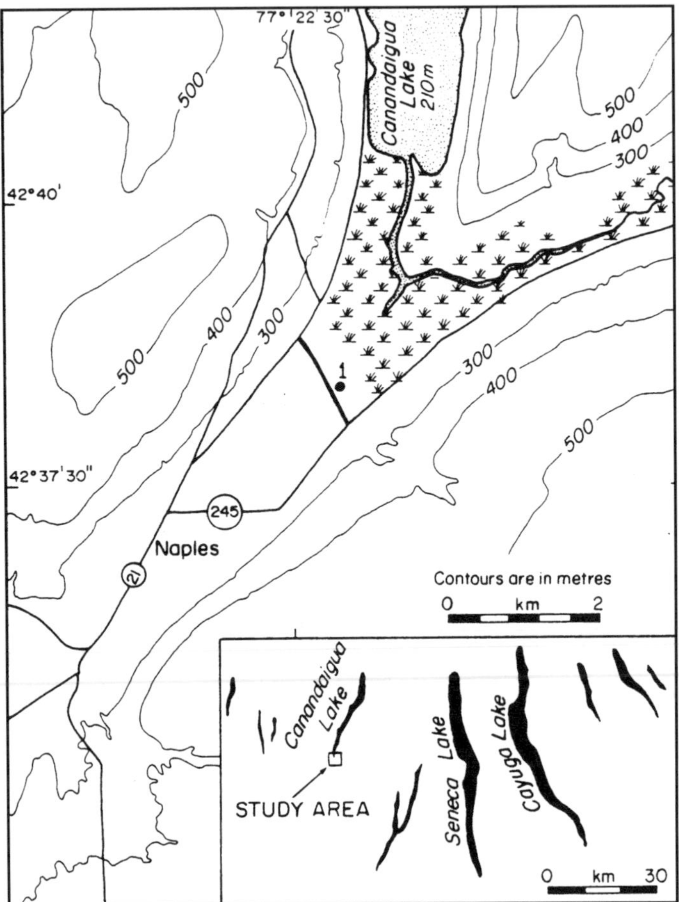

Figure 1. Location of the borehole (labeled 1) and seismic reflection line (bold solid line) in the dry valley south of Canandaigua Lake. Inset shows location of the study area and Canandaigua Lake relative to the other Finger Lakes. (Adpated from H. T. Mullins and R. W. Wellner, personal communication.)

Subsurface stratigraphy

The general subsurface stratigraphy of the study area is based on a qualitative correlation of an onshore weight-drop seismic reflection profile with gross drillcore results (Fig. 2) (see Wellner et al., this volume). Four first-order seismic stratigraphic units have been identified beneath the dry valley south of Canandaigua Lake (Fig. 2), as opposed to the six sequences on the higher resolution reflection data from the lake itself: (1) overlying bedrock, a basal sand and gravel unit that correlates with the Valley Heads Moraine (14,400 to 14,100 ^{14}C yr B.P.) (see Wellner et al., this volume) and with sequence I of the offshore stratigraphy; (2) a massive to bedded glaciolacustrine silt and sand, approximately 90 m thick and with gravel layers present, that correlates with offshore sequences II and IV; (3) a 15- to 20-m-thick alluvial sand and gravel aquifer with artesian (overpressured) water that correlates with offshore sequence V; and (4) a surficial lacustrine sequence of interbedded clay and peat, that is about 10 m thick.

On the basis of sedimentologic analysis, the subsurface stratigraphy south of Canandaigua Lake may instead consist of five first-order units and six second-order subunits (Fig. 3) (see also Wellner et al., this volume). A basal sand and gravel (unit 1), which terminated drilling, is overlain by massive clay and silt devoid of dropstones (unit 2a) and capped by a coarse sand and gravel layer (unit 2b). Unit 3, identified on reflection records from Canandaigua Lake (see Mullins et al., this volume) does not continue beneath the dry valley south of the lake. Unit 4 consists of four second-order layers that alternate between massive clays with dropstones (units 4a and 4c) and well-bedded silt and sand (units 4b and 4d). Unit 5 is, as before, an alluvial sand and gravel layer, capped by the fine-grained clay and peat of unit 6.

DOWNHOLE GEOPHYSICAL LOGGING

Methods

The logging was carried out using downhole resistivity tools in tandem with a simple natural gamma-ray (γ-ray) emission counting tool. Two different resistivity electrode spacings were used: 0.406 m (16 in.) and 1.626 m (64 in.). The shorter spacing yielded finer scale results with shallow penetration; deeper penetration into the formation beyond the borehole wall was obtained with the larger electrode spacing, but there was an associated loss of resolution. A steel casing was required for the upper 26.8 m of the hole, to seal the artesian sand and gravel aquifer (unit 5), and thus there are no resistivity results for that portion of the borehole. The tools were lowered to refusal (approximately 112 m) near the bottom of the hole.

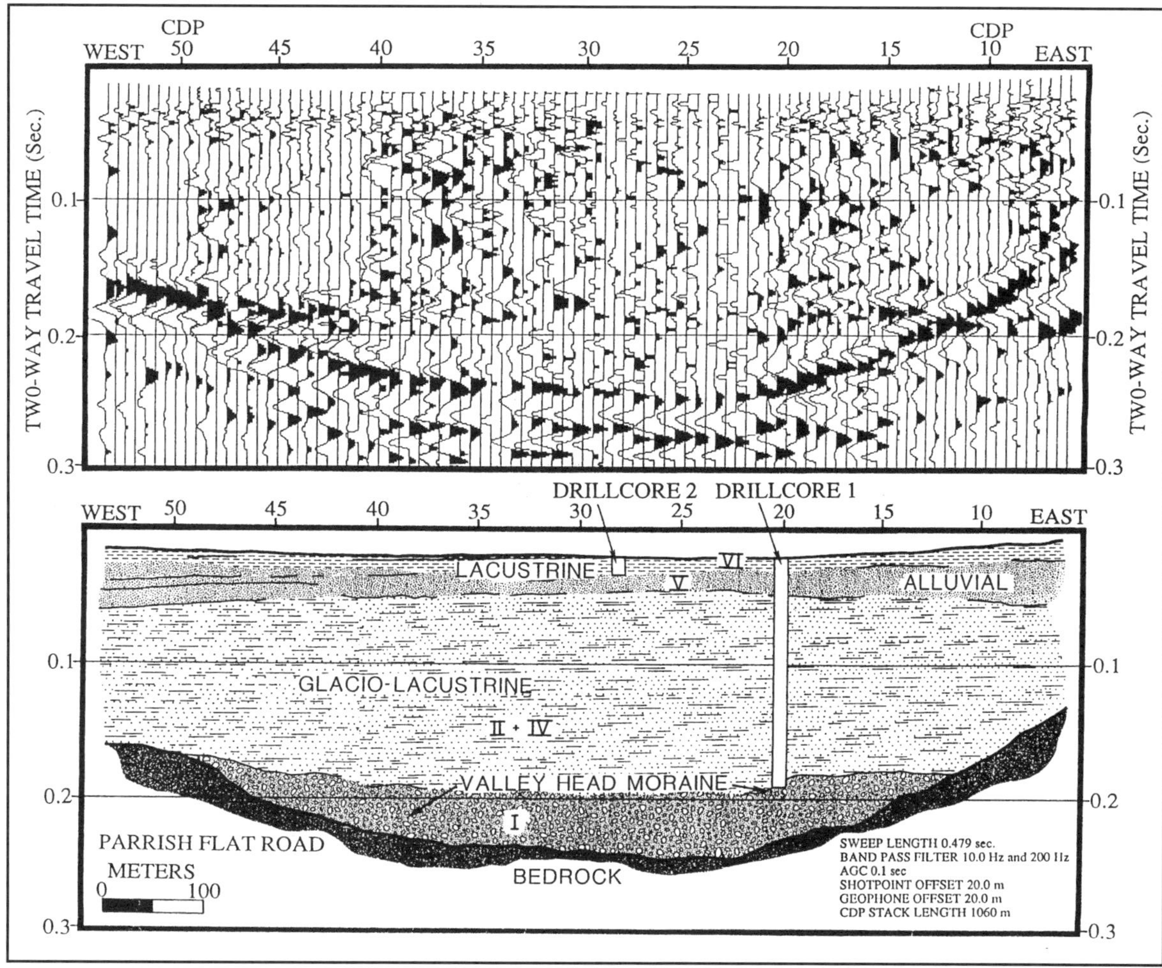

Figure 2. Onshore shallow seismic reflection profile taken along Parrish Flat Road (top), and the lithologic and seismic stratigraphic interpretation (bottom). The downhole logging and VSP were carried out in drillcore 1, near Parrish Flat Road. (Modified from Wellner et al., this volume.)

There are no results for the 1.6-m resistivity at the very base of the hole, because the long electrode spacing prevents logging of the basal few meters. Logging was carried out while the tools were raised to the surface, so that the cable was under tension, thus allowing a consistent sampling rate per second to be transformed directly into a sampling interval in meters.

The logs were acquired using 0.5- and 3-sec sampling windows for the natural gamma tool. The γ-ray activity results were comparable, but the 0.5-sec data were quite noisy due to the low count rate. The statistical variability is reduced when a larger time window is used, because some of the random fluctuations in the count rate are averaged out. However, even the 3-sec count data were noisy, and a further 11-point tapered running mean was used to yield the results that were ultimately used for the analysis and interpretation. To check the consistency of the results, all of the analyses were carried out for both the filtered and unfiltered 3-sec natural γ-ray emission data. The results were identical, except for the associated slight increases in the standard deviations and decreases in the correlation coefficients. Here we present only the final results using the smoothed γ-ray emission data, although we do present the raw 3-sec window data in Figure 3.

The γ-ray emission count in sediments is largely a function of the clay content. Similarly, the electrical conductivity (the inverse of the resistivity) depends on the water content, water quality, and the clay content. In general, the conductivity can be separated into a contribution due to the water and a clay contribution, and would usually be of the form:

$$1/\rho = \sigma = \sigma_w \phi^n + \sigma_c , \tag{1}$$

where ρ is the formation electrical resistivity, σ is the formation electrical conductivity, σ_w is the pore fluid (water) conductivity, σ_c is the clay contribution, and ϕ is the porosity (e.g., McNeill, 1990). The exponent n is sometimes called the "shape factor" (Jackson et al., 1978; McNeill, 1990) or "cementation exponent" (Labo, 1986; Ellis, 1987), and has a value that ranges from approximately 1.4 for pure spherical grain sands to 2 for clays. Transforming the resistivity to the conductivity,

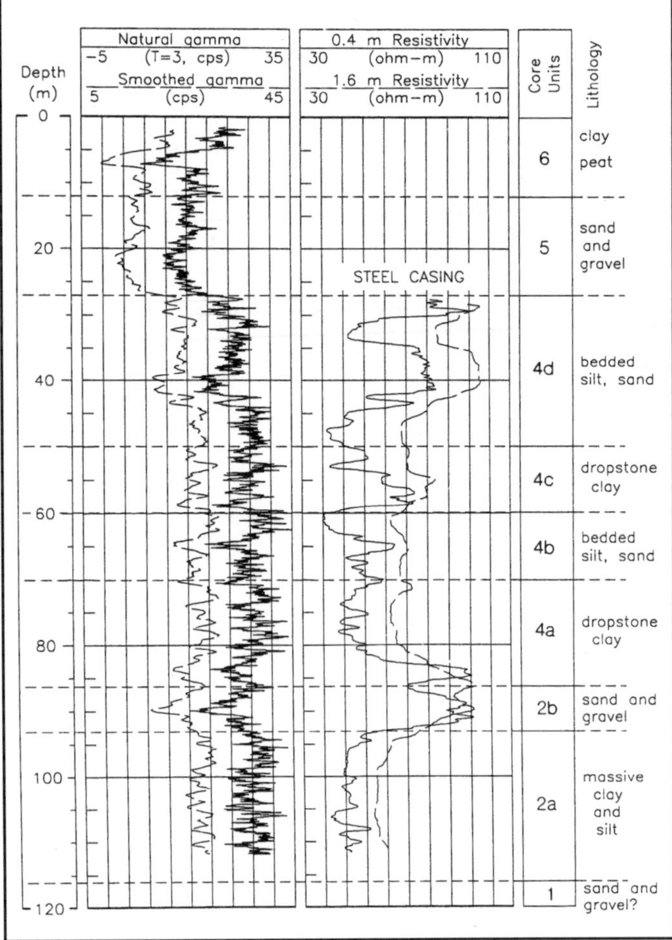

Figure 3. The raw geophysical logs, plotted alongside the lithology interpreted from the drilling results. The natural γ-ray activity (in counts per second [cps]) for a 3-sec window is shown both in raw (solid) and smoothed form (dashed, far left). The raw 0.4-m (solid) and 1.6-m (dashed) resistivities are in column 2 (center). The boundaries of the lithologic units as defined from the coring are delineated by the horizontal dashed lines, and the lithologic units are noted to the right of the logging data. While there is general agreement between the logging data and the previously defined core lithologies, the boundaries do not correspond exactly.

by taking its inverse, we can expect both the γ-ray activity and the conductivity to increase in the presence of clay, and to decrease in sand layers. The conductivity is preferred, over the resistivity, because of its simpler form, as represented in (1), and because it responds to an increase in clay content in a way that parallels the γ-ray activity. By contrast, the resistivity depends in a more complex way on a combination of the water content, water quality, and clay content.

Principal components analysis

In addition to calculating the electrical conductivities as the inverse of the electrical resistivities, we carried out a principal components analysis (PCA). PCA is akin to a translation and rotation of the data coordinates from some standard set, in this case the γ-ray activity and electrical conductivity, to a new set that better represents the major data trends (Fig. 4). These new axes are linear combinations of the old axes. The first principal component, PC 1, is then aligned in the direction of maximum variance, and the residual, PC 2, is aligned perpendicular to PC 1. Such analyses can be used to distinguish lithologies (Kassenaar, 1989, 1992), and have been used to identify clay layers and ground-water aquifers (Kassenaar, 1992).

The major trends in conductivity and γ-ray activity (Fig. 4) suggest a strong influence due to variations in clay content. The first principal component, PC 1, accounts for 82% of the variance, and is aligned so that the γ-ray activity increases as the conductivity increases. PC 1 correlates well with the core lithologies (Fig. 5), and this good correlation reinforces our suggestion that PC 1 is distinguishing between clay-rich and sand-rich layers. The PC logs are calculated in standard deviation units, i.e., the data have the means removed and are divided by the standard deviations in the directions of the respective principal components.

The second principal component, PC 2 (the residual component), accounts for the remainder of the variance in the data, but does not appear to have a clear lithologic interpretation (Fig. 5). The few peaks that seem to be anomalous also appear

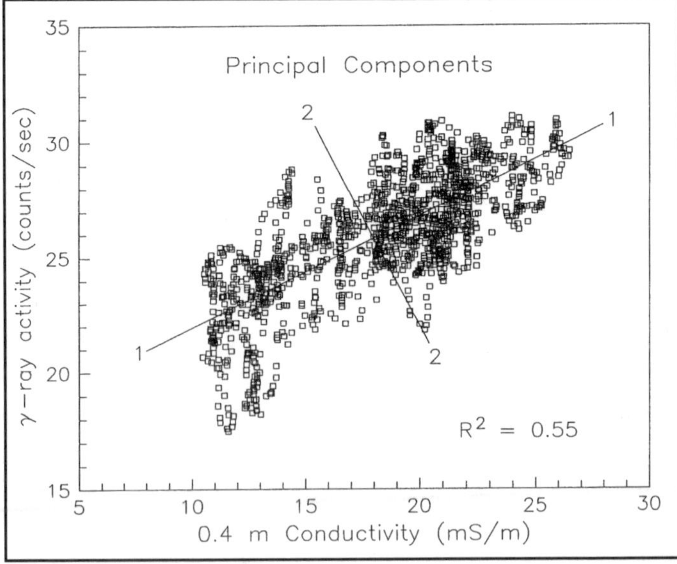

Figure 4. Correlation of γ-ray activity and 0.4-m conductivity, showing the principal components. The first principal component, indicated by the long axis along the trend of the data points (labeled 1), accounts for 82% of the variance in the γ-ray activity-conductivity crossplot. The remainder of the variance is contained in the second principal component, indicated by the short axis (2) perpendicular to the first principal component. Principal component analysis simply involves a rotation and translation of the original data axes, γ-ray activity and conductivity, into two new orthogonal axes, 1 and 2. The new coordinates are normalized; that is, the data values are divided by the standard deviation in each axis direction, so that the data in each direction have 0 mean and a standard deviation of 1.

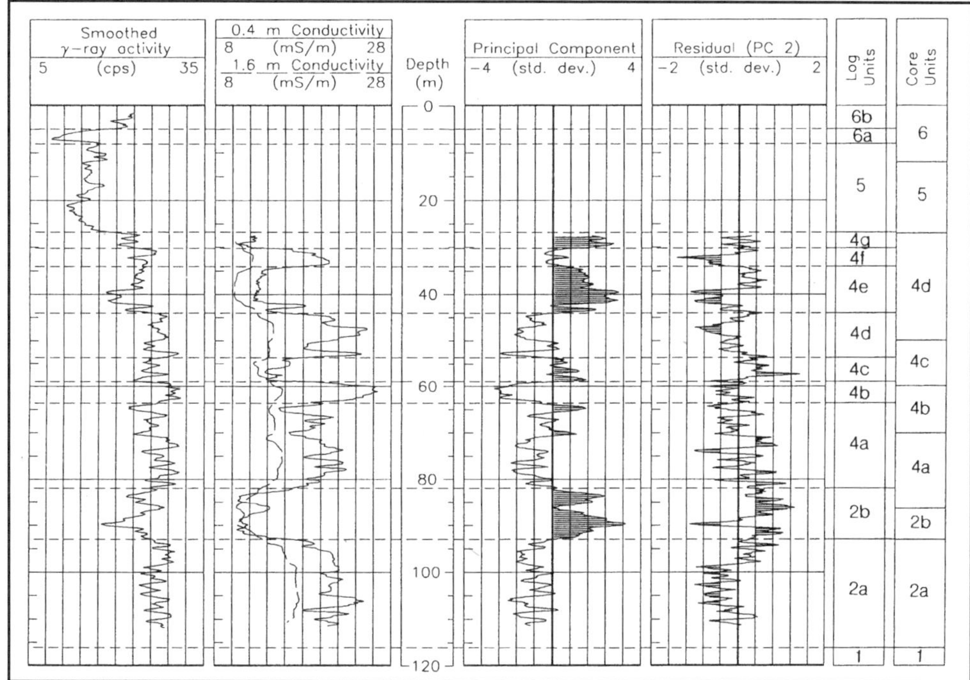

Figure 5. Smoothed γ-ray activity (far left), calculated electrical conductivity (center left), and principal components (center right) versus depth. The PCA (principal components analaysis) results are plotted in standard deviation units. The first principal component (PC 1) is labeled Principal component, and PC 2 is labelled Residual, because it represents the variance left over from PC 1. PC 1 appears to differentiate between layers with higher sand contents (shaded) versus higher silt and clay contents (unshaded). High residual values (shaded) appear to emphasize these characteristics, but with the opposite sign. New lithologic units have been defined using the electrofacies, and are indicated in the next-to-last column on the right (Log Units), beside the previously defined units (Core Units). The boundaries of the Log Units are delineated by the horizontal dashed lines. Note the general background increase in the γ-ray activity and the conductivity with depth, indicating a decrease in grain size and increase in clay content with depth. The conductivity trend is best seen in the large electrode separation data, for the 1.6-m conductivity.

to be aligned with zones already distinguished by high (sand) or low (clay) PC 1 values. PC 2 is not used in our interpretation because it appears to contribute little additional information.

Comparison of geophysical and drillcore results

The calculated electrical conductivities, the smoothed γ-ray activity, and the principal components are presented in Figure 5. High positive values for the first principal component (shown shaded in Fig. 5) appear to correlate with sandier units, while low (large negative) values are indicative of high silt and clay contents. While the correlation between the core lithologies and the logging data is generally good, the core unit boundaries are not directly aligned with changes in the logs, and additional second-order units can be identified. The differences are due to the incomplete core recovery versus the continuous nature of the downhole geophysical data. A set of units and subunits, different from the core units, can be defined based on the logging results (Fig. 5), i.e., electrofacies (Ellis, 1987; Labo, 1986), although we have kept the same basic framework as used for the core lithologic units.

Starting from the base of the hole, electrofacies unit 1 is the same as core unit 1. We were not able to sample that section of the hole but do see evidence for it in both the surface seismic results and in the VSP, which will be discussed in a later section. Unit 1 is, as noted earlier, composed of sandy gravels of the Valley Heads Moraine. Electrofacies 2a is the same as core unit 2a, which is a massive silt and clay layer. The γ-ray activity and the conductivity are both high, and PC 1 is low. Overlying unit 2a is unit 2b, which is similar to core unit 2b, but thicker. Unit 2b is primarily a sandy gravel, with a thin bed of silt and clay at about 87 m. Unit 2b is distinguished from the bounding units by its anomalously low conductivity and somewhat lower γ-ray activity. The labeling of units is, to some degree, arbitrary; the separation of units 2a and 2b from the overlying unit 4 is based on changes in trends and in thicknesses of interbedded layers. While subunits 2a and 2b are distinctly different and the boundary between units 2a and 2b is sharp, they are grouped together to be consistent with core unit 2 and seismic stratigraphic sequence II as defined elsewhere (Wellner et al., this volume; Mullins et al., this volume).

Unit 4, overall, is composed of a series of interbedded sands and silts or silty clays. The γ-ray activity and the conductivity generally increase downhole, which indicates that the clay content is increasing with depth in the unit. The trend in the conductivity is best noted in the 1.6-m conductivity, which, because of its large electrode separation, is less sensitive to thin beds. There is, therefore, a general undulating trend through unit 4, with a sudden change into unit 2b. Unit 4 has been subdivided into seven subunits. The boundaries are defined where two conditions are met: where there are sharp, distinct changes in the geophysical logs, particularly the principal component log; and where such sudden variations do not return to previous levels within a couple of meters. Thus, subunit 4a is composed of a series of alternating beds, but with a general, subtle coarsening-upward trend, and moderate conductivity (Fig. 5). Subunit 4b, by contrast, is a more homogeneous, clay-rich layer, with high γ-ray activity and conductivity. Subunit 4c is a sandier layer that contains two identifiable thin beds with elevated clay content, at depths of 55 and 57 m. Subunit 4d is principally a silt layer; the γ-ray response is not quite as high as that for subunit 4b. There is a distinct sand layer within 4d, at a depth of 50 to 52 m, but the log values are similar above and below this layer. Subunits 4e, 4f, and 4g are alternating layers of sand and silt. Subunit 4e is sand and sandy silt with a thin but distinctive layer of silt at 42- to 43-m depth; subunit 4f is silt with a thin sand layer at 37-m depth; and subunit 4g is a sand layer. Taken as a whole, subunits 4e, 4f, and 4g fit into a trend of increasing γ-ray activity and increasing conductivity, i.e., increasing clay content and decreasing grain size, with depth in unit 4.

Units 5 and 6 are entirely within the steel casing, and thus cannot be defined on the basis of a complete set of logging data, because no electrical logs are available. Based solely on the γ-ray activity, electrofacies unit 5 is the same as core unit 5, a sand and gravel layer, but extends from depths of about 26.5 to 8 m. Overlying unit 5 is unit 6, which can be subdivided into a peat layer, subunit 6a (5 to 8 m), and a surficial lacustrine clay, subunit 6b (5 m to surface).

Effects of interbedding

On close inspection of the electrical logs (Figs. 3 and 5), it is apparent that the 1.6-m conductivity is generally less than the 0.4-m conductivity, and to a large extent defines a minimum value. Only in unit 2b are the two logs approximately equal. In subunits 4a and 4c, the 1.6-m conductivity is not much less than the values for the shorter electrode spacing. That is, the two conductivities are equal or nearly equal only within sand layers, with little evidence of interbedding. Conversely, the differences between the 0.4- and 1.6-m conductivities in the other units and subunits are likely a result of interbedding of sand and silt layers. The larger electrode spacing of the 1.6-m electric log yields an average of the electrical properties over a greater depth range than the 0.4-m electric log. The average is not a simple arithmetic average, but rather is taken as the average along the direction of the borehole, perpendicular to the bedding, and is lower than the lowest readings obtained from the 0.4-m electric log. The near equality of the two electric logs in unit 2b suggests that the log differences are not due to a calibration problem, but are evidence for real structure.

Both the 0.4- and 1.6-m resistivities represent averages taken perpendicular to the bedding, as opposed to an average along the bedding direction, as briefly discussed further in appendix 1. In principle, we have no knowledge of the horizontal properties, parallel to bedding. The 0.4-m resistivity, however, is taken at a smaller interval, and thus represents an average over fewer layers, which should be closer to the actual formation resistivity. While it is not known how finely bedded the formation is, nonetheless, if the electrical properties of each layer are identical, there is no anisotropy, and the measured 0.4- and 1.6-m electric logs are identical. We can, in the minimum, define a *bedding index*, which is calculated as the ratio of the 0.4-m conductivity to the 1.6-m conductivity (Fig. 6). The bedding index provides some measure of the amount of bedding and the degree to which the electrical properties differ between alternate layers. The largest values are obtained in those layers that have the highest apparent silt and clay contents, as indicated by the γ-ray activity and PC 1; that is, the platy grained clays tend to have higher degrees of anisotropy, and thus have higher bedding indices. By contrast, sand grains tend to be more spherical, and thus have lower bedding indices. The bedding index results reinforce our interpretation of the principal component response, although subtle differences do exist. For example, subunit 4f does not have a significant principal component response, but does have a large bedding index. This suggests that subunit 4f is composed of a number of thin silt and sand layers that cannot be individually resolved.

VERTICAL SEISMIC PROFILING

Methods

A vertical seismic profile (VSP) was collected (Fig. 7) to calibrate surface seismic reflection profiles and to correlate the reflectors with features observed on the borehole logs. A set of 11 hydrophones was placed in the Canandaigua borehole, with the topmost hydrophone at a depth of 1 m. A seismic "shotgun" source was positioned 10 and 20 m from the borehole; the best results were obtained using the 10-m offset. For each shot, the direct arrival and any subsequent reflections were recorded on a 12-channel digital seismograph, and the data were stored on diskette. The hydrophone string was then lowered to the next depth interval for the next shot. This procedure was repeated to refusal, at 99 m. Because any given shot hole is quickly turned into a subsurface cavity that has lost its coupling with the surroundings, a series of shot holes was used, arranged radially about the borehole at constant distances from the borehole.

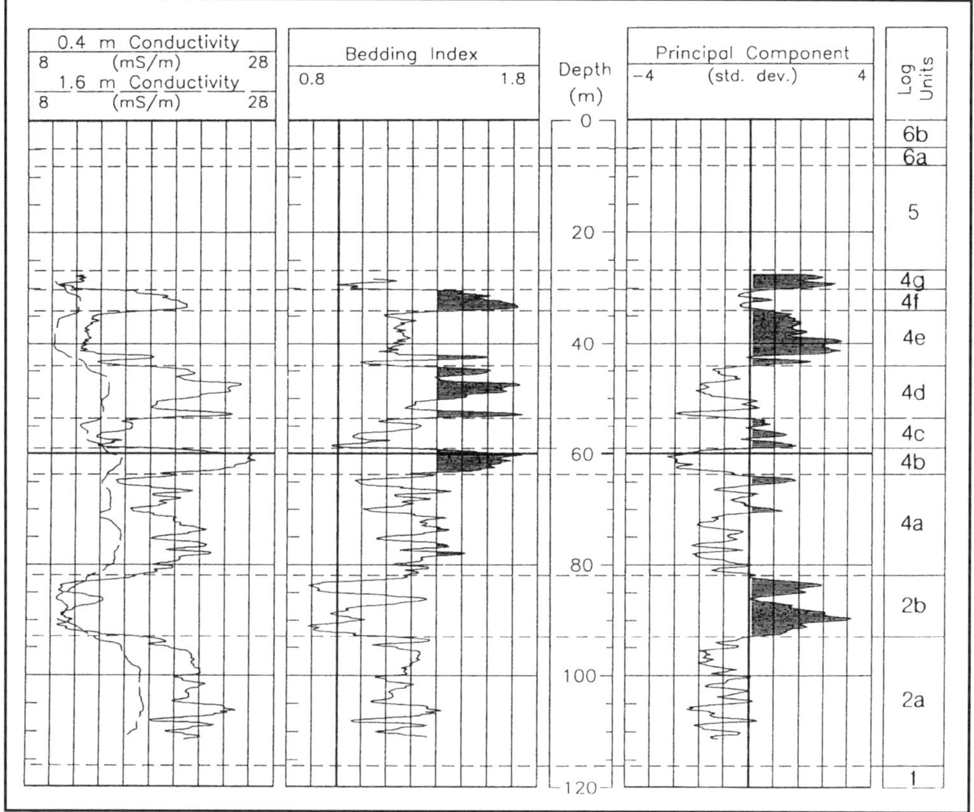

Figure 6. Comparison of the conductivity logs and the first principal component with the bedding index. See text for a detailed discussion.

The data were copied from the diskettes to a 386 microcomputer for analysis using a seismic processing package. The raw VSP record has both the direct, downgoing waves recorded, and the upgoing energy that has been reflected at boundaries where the impedance contrast (density and/or acoustic velocity) changes significantly within a short depth interval. The raw VSP section is dominated by the downgoing energy (Fig. 8). The VSP energy can be separated into downgoing and upgoing components by transforming the raw data from the time-depth domain (Fig. 8) into the frequency-wave number, *f-k*, domain (Fig. 9). In the *f-k* domain, the downgoing energy is contained entirely in the negative wave number quadrant, i.e., for $k < 0$, and the upgoing energy is in the positive wave number quadrant, i.e., $k > 0$. The data are filtered to suppress one side or the other in order to isolate the energy to be examined.

Results and correlation of logging, VSP, and shallow seismic results

An advantage of the *f-k* domain is that arrivals for a simple reflector, or for the direct arrival, show up along a discrete band centered about a line (e.g., Hatton et al., 1986). For example, the direct arrivals fall along the line indicated in Figure 9. This simple feature arises from the relationship between velocity, V, frequency, f, and wavelength, λ, in wave propagation theory (e.g., Telford et al., 1990):

$$V = \lambda f. \tag{4a}$$

Since the wavelength, λ, is the inverse of the wave number, k, then we have:

$$V = f / k \quad \text{or} \quad f = k V. \tag{4b}$$

Equation (4b) defines a straight line relating frequency to wave number, with a slope equal to the velocity, V. The straight line relationship derived from the direct arrival in the *f-k* domain yields a velocity of 1,669 m/sec. The line in the *f-k* domain is actually slightly curved, and the energy is distributed across both frequency and wave number.

An average velocity above a reflecting horizon can be obtained directly, by simple dividing the depth of a particular hydrophone by the time of the first direct arrival at that hydrophone (Fig. 10). The velocity values are lower than the *f-k* value because the direct calculation represents an average for all layers that overlie a given hydrophone. The velocity in the near-surface vadose (unsaturated) zone is quite low, but increases quickly with increasing water saturation, and the average velocity consequently increases quickly with depth near the top of the borehole.

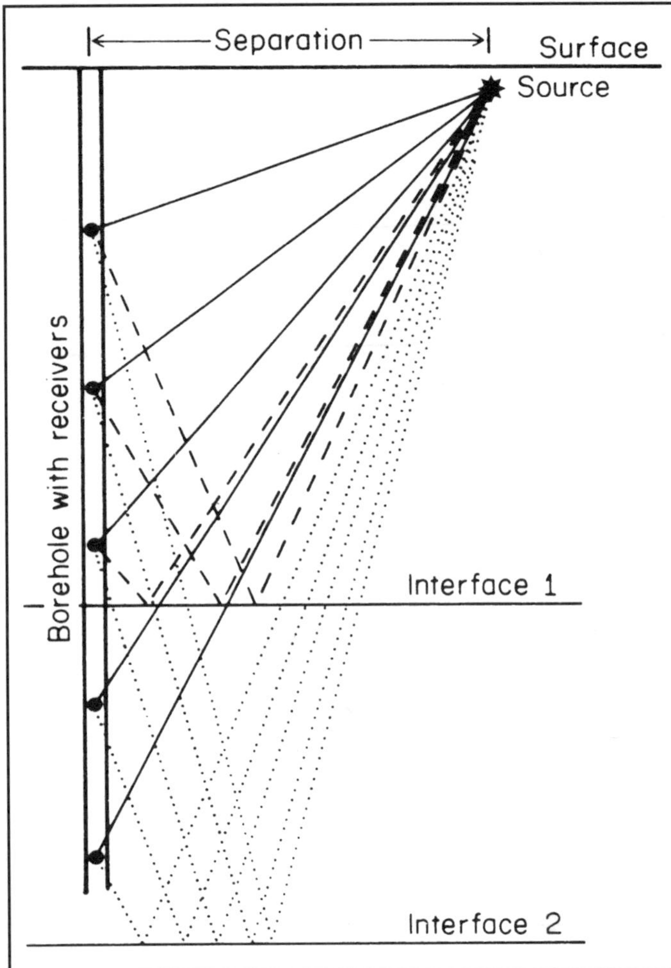

Figure 7. Schematic drawing illustrating the basic principles of vertical seismic profiling (VSP). A shot is placed some distance from a borehole containing a set of seismic receivers. The direct, downgoing arrival is recorded at all receivers (solid lines) at progressively later times at greater depths. Interfaces where there are physical property contrasts will reflect some of the energy back up to the borehole receivers, e.g., interface 1 (dashed lines). Reflections can arise as well from boundaries that lie below the base of the borehole, e.g., interface 2 (dotted lines).

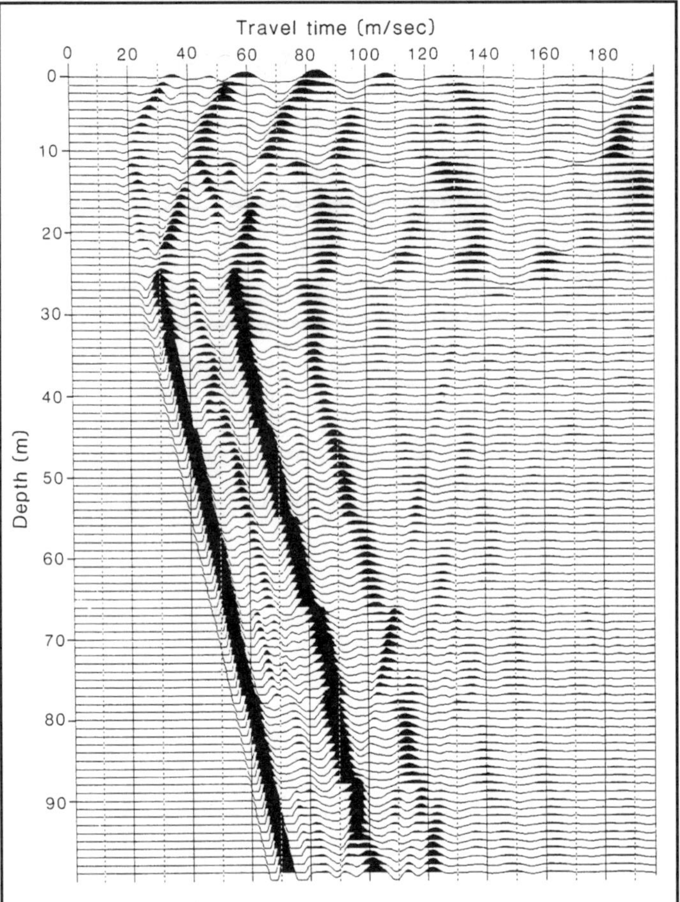

Figure 8. An example of a raw VSP section, using the record for the 10-m shot-borehole offset. Note the dominance of the downgoing energy. Few upgoing events can be clearly identified.

The upgoing VSP energy, once isolated, can be compared and correlated with the logging results (Fig. 11). The source of each reflector may thus be identified, and we can determine the (surface) two-way travel time for a given reflector. There are 11 large-amplitude reflectors, and a number of low-amplitude reflectors. Some of the reflectors can be identified as couplets or pairs of reflectors, one positive and the other negative, which arise from the top and bottom of thinner beds, such as the events labeled A, C, and D (Fig. 11). Many of the units are composed of interbedded layers; a few of these intermediate beds give rise to reflections, albeit of lower amplitude, for example events B, H, I, and J. Most of the reflections, however, arise from the unit and subunit boundaries: A arises from the top and bottom of subunit 4f; the C and D couplets lie at the bases of subunits 4d and 4c, respectively; E originates from the base of subunit 4b; G originates from the boundary between units 2 and 4; and K is the reflection from the top of the Valley Heads Moraine. L and M are interpreted as originating from within the morainal unit, and are of undetermined origin. Events N and P may be related to basement; these events are considered in more detail when the correlation with the surface seismic segment is discussed.

Some of the unit and subunit boundaries do not give rise to reflections. The boundaries between subunits 4e and 4d, and between subunits 2b and 2a, must be transitional, because they do not generate any notable reflections. The nature of the transition may, however, change with position, and so the boundaries may generate patchy reflections at other locations. Some of the events that are observed may, similarly, change with position, so that the events are patchy and uneven. Those reflectors within beds, in particular B, H, I, J, L, and M, will probably be irregular in appearance on any surface seismic reflection record.

The VSP reflections are, in turn, correlated with the surface seismic reflection record (Fig. 12). Because the VSP section has been processed, and the reflectors "flattened" to correspond with the two-way travel time, then the distortion caused by the steel casing is less of a problem. Shallow reflec-

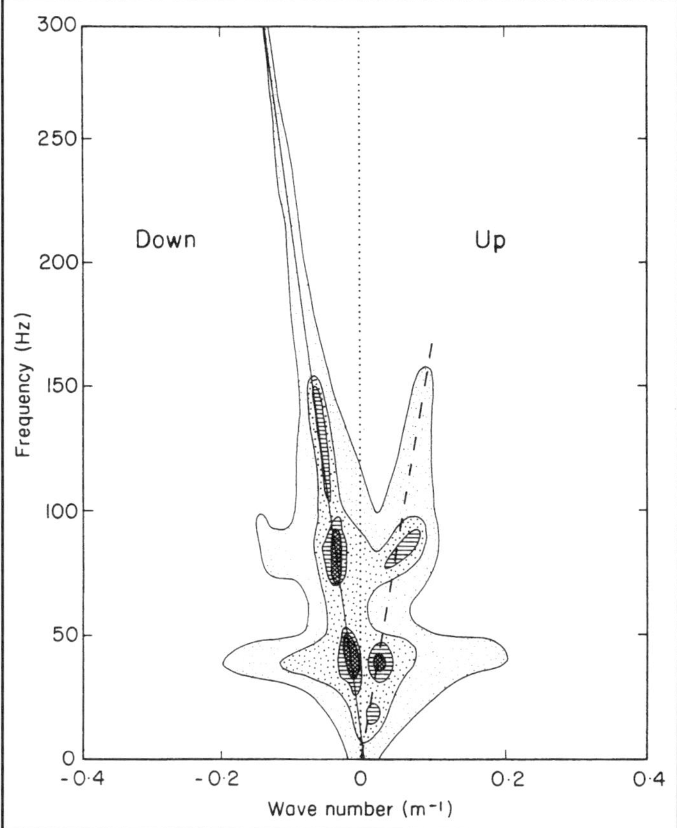

Figure 9. Spectral analysis separates the upgoing energy (to the right) from the downgoing energy (to the left), $k > 0$ and $k < 0$, respectively. The smoothed contour intervals are unevenly spaced: the outermost contour encloses those portions of the *f-k* plot where the amplitude is 10% of the peak amplitude or greater (light stippling). The next level is approximately 30% or greater (heavy stippling), the third is 60% or greater (horizontal shading), and the innermost level is 90% or greater (cross-hatched). There is an approximate trend in the upgoing energy, as indicated by the dashed line, but the energy distribution is rather lobate and not simple. While the downgoing energy distribution is also lobate, there is a significant portion of the energy centered about a line with an apparent velocity of 1669 m/sec.

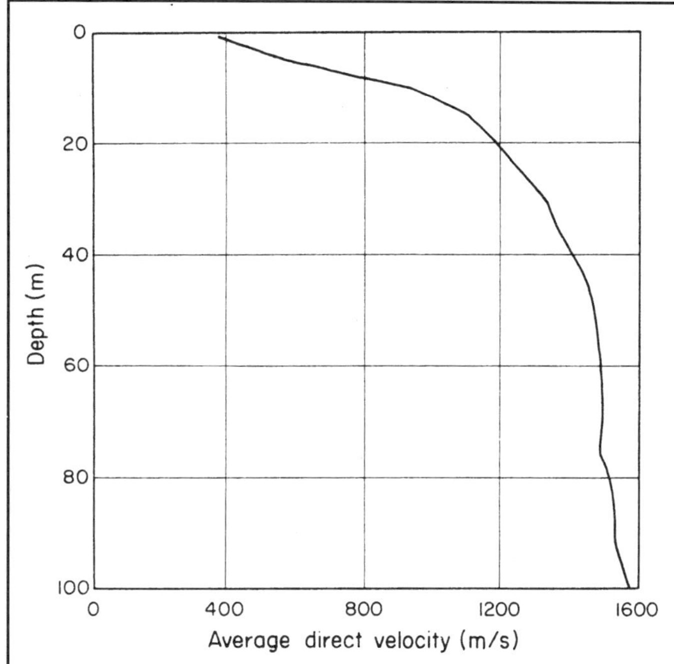

Figure 10. Average direct velocity versus depth, as calculated from the time of the first arrival for a given hydrophone depth. This velocity is an average over all of the layers above the particular hydrophone, and is thus less than the velocity in any specific layer. Note the sharp increase in velocity in the near surface, as the signal travels from the vadose (unsaturated) zone through the water table down into the saturated zone.

tors cannot be identified. This is less of a concern since there are few such reflectors, and the core recovery and high resolution natural γ-ray activity log in the upper 27 m allow us to identify possible reflectors with some confidence. The sources of the deepest reflectors may be less obvious. The correlation of the VSP section with a segment of the seismic reflection section is good (Fig. 12), and most events can be identified in both sets of data (Table 1). The shallow events, A through G, occur at approximately the same times in both sections. As expected, reflectors B, H, I, and J are difficult to clearly identify. The reflector from the top of the Valley Heads Moraine (unit 1), K, lies deeper in the surface seismic section, which suggests a slight thickening of unit 2 between the borehole and the location of the surface seismic section. This trend would be a slight deviation from the regional pattern, since the Valley Heads Moraine is present at the surface a few kilometers to the south (Wellner et al., this volume). The relative time delay between the top and bottom of unit 1 is difficult to determine for the surface seismic section, because the basement reflection is not recorded on the hammer seismic record acquired for comparison with the VSP and borehole logging data. If basement is expected to lie at about 200-msec, two-way travel time, on the surface seismic segment shown (Fig. 12), then N may be the basement reflector; if, on the other hand, we use the delay time between the deeper VSP reflectors and the corresponding events identified on the seismic section, then event P may instead be the top of basement. However, N yields a stronger reflection than P; we suggest that N represents the top of the basement, and P is therefore a weak intrabasement reflector, which would likely be patchy and intermittent in character if it were recorded in the surface seismic section.

CONCLUSIONS

The lithologic units in the borehole located south of Lake Canandaigua have been revised and refined using the results of downhole geophysical logging. The logging units are similar to those defined using drillcores, but the boundaries have been modified, and unit 4, which is primarily composed of interbedded sands and silts, has been subdivided into seven subunits,

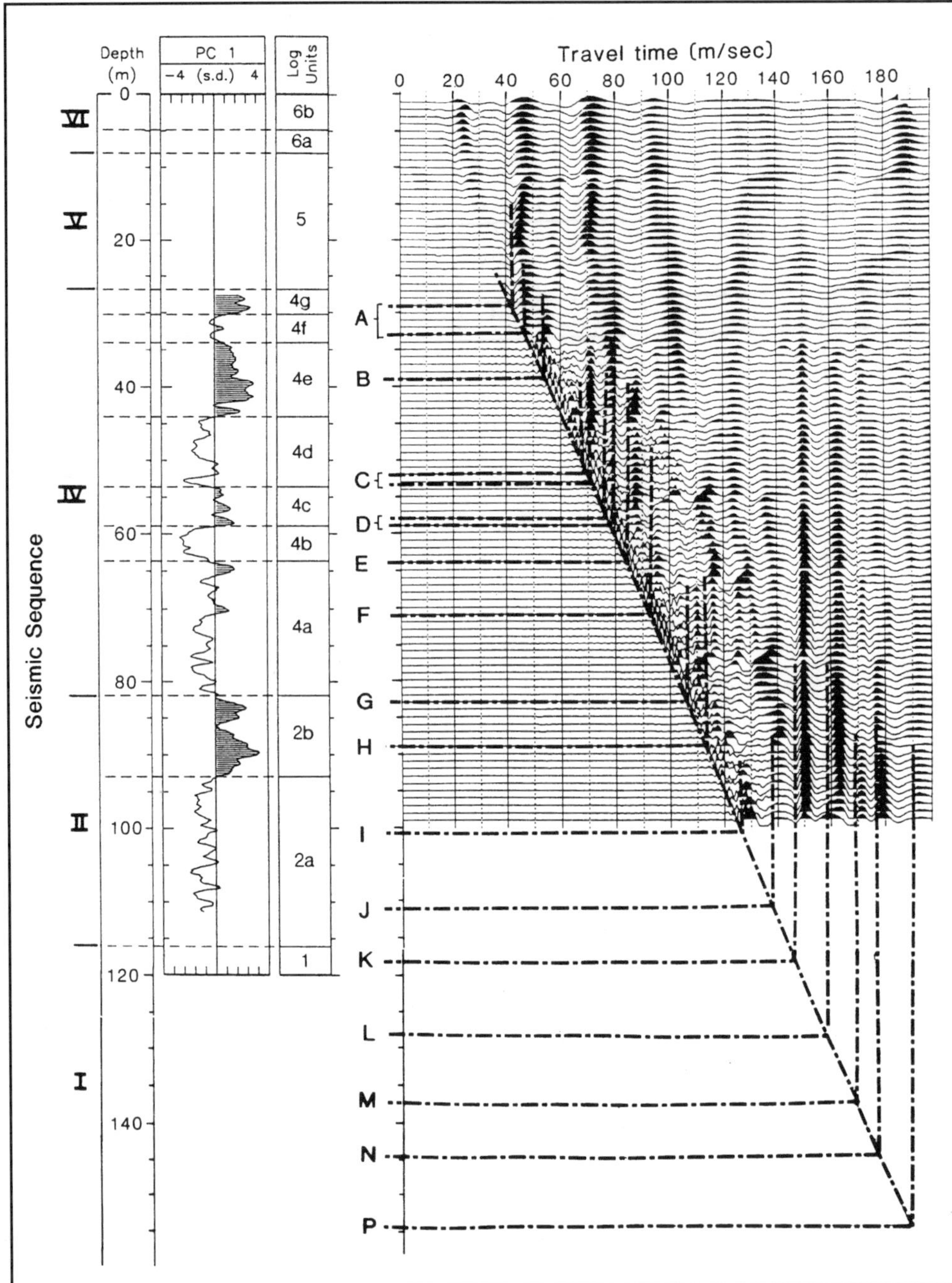

Figure 11. Comparison of the VSP upgoing energy, processed so that the reflective events are aligned in time to coincide with the two-way travel time, plotted using the same depth scale alongside the first principal component log and the electrofacies units. The reflectors are labeled from just below the steel casing down; all times are approximate. The near surface event (not labeled) is the direct arrival, not a reflection, in the vadose zone. Event A (45 msec) is a negative-positive couplet associated with the top and bottom boundaries of silty clay subunit 4f (30 and 34 m, respectively). B (52 msec) is a weak reflector from the top of the sandy bed within subunit 4f (38 m); the larger value of the PC 1 log is due largely to lower γ-ray activity. C (70 msec) is a negative-positive couplet from the top and bottom of the clay-rich bed at the base of subunit 4d (52 and 54 m); there is also a very weak reflection (unlabeled) from the top of the silt bed that overlies the clay-rich bed. D (80 msec) is a positive-negative pair associated with the top and bottom of the sandy bed at the base of subunit 4c (57 and 58 m). E (85 msec) arises from the base of subunit 4b (64 msec). F (92 msec) is a weak reflection from within subunit 4a (~70 m). There may be other, very weak reflections from the interior of subunit 4a, but these have not been indicated. The G reflector (110 msec) is from the boundary between units 2 and 4, and H (115 msec) appears to originate from the sand layer within subunit 2b, at a depth of approximately 88 m. The boundary between subunits 2a and 2b is transitional, and does not give rise to a VSP reflector. Events I and J are weak reflectors from beds within subunit 2a. K (150 msec) is a strong, distinct reflector from the top of unit 1, the Valley Heads Moraine; because the VSP reached only to a depth of 99 m, the depth of event K, 122 m as estimated from the VSP, is not exact, but is close to the value of 118 m determined from drilling. Events L (160 msec, ~130 m) and M (170 msec, ~138 m) appear to come from within unit 1, based on the surface seismic reflection results. Finally, events N and P (175 msec, 142 m, and 185 msec, 155 m) may be within unit 1 or may come from the top of basement.

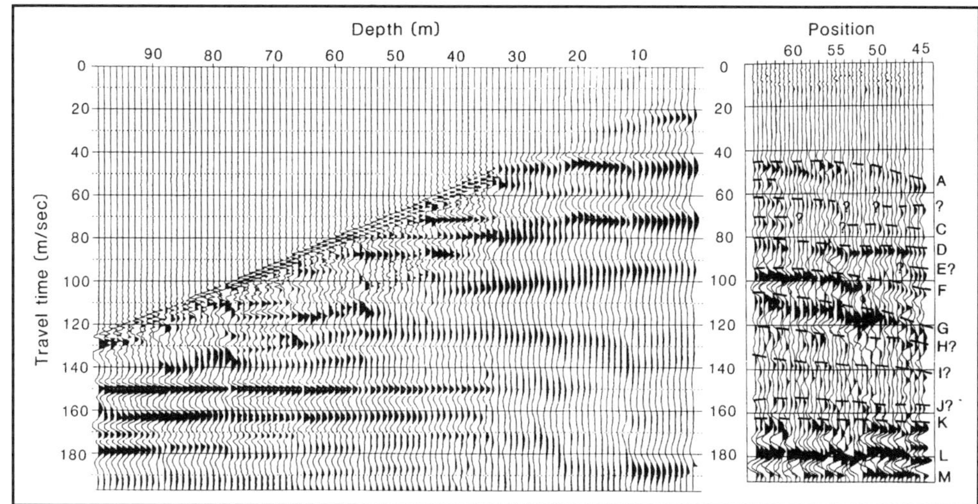

Figure 12. The processed VSP of Figure 11, turned on edge to show the correlation with the surface seismic reflectors for a segment of the surface seismic reflection line that is relatively close to the borehole; the shot position is marked. The correspondence between VSP and surface seismic reflectors will not be exact for a variety of reasons: the frequency content, and hence the resolution, of the two sections is different; the VSP will record weak reflectors that may not be recorded at the surface; the surface seismic reflection line is offset from the drillhole; and the depths of lithologic boundaries will vary locally. Nonetheless, the correlation of the VSP reflectors with the seismic section is very good, as indicated along the right hand side and summarized in Table 1, using the labels as assigned in Figure 11. The origin of the reflector between A and C is unknown; there are in fact two events on the left side of the surface seismic section and one on the right-hand side, and they could arise either from B or from the weak event above B, or both. Similarly, events E and J may or may not be present in the surface seismic section. Event N does not appear to be present, lying below the 198-msec sweep of the surface seismic section; basement is expected to lie at about 200-msec two-way travel time, and thus N is preliminarily interpreted to be the top of basement.

instead of four. The results of a principal component analysis allow for a more direct and objective delineation of the units, in response to the sand versus silt or clay contents of the units and subunits. The electrical properties indicate that there is a significant amount of interbedding, and in some layers the degree of bedding must be quite high. We have defined a *bedding index*, the ratio of the short- to long-electrode spacing electrical conductivities, to try to isolate zones with a high number of thin, unresolvable beds. The unit and subunit boundaries defined using the logging results correlate well with the reflectors identified in the VSP results, although some boundaries yielded no reflection events, while some reflections appeared to originate from within a unit or subunit. The correlation with a portion of a surface seismic section is good; most of the reflectors in the VSP section can be recognized in the surface seismic section.

The results indicate the presence of sequences of finely bedded, possibly even laminated, sediments, which may suggest annual to decadal variations in sediment deposition at various times in the past. The apparent thickening of unit 2 between the borehole and the seismic section is contrary to the observed regional pattern, but the difference is small and such variations in the topography of the unit boundaries are not unusual. There are reflectors from within the Valley Heads Moraine; the sources of the reflections are not known. The basement reflector is not definitively identified, but based on the strength of the reflector and the relative time delays between the surface seismic and VSP events, a reflector at a depth of approximately 142 m may be the top of basement. If that is the case, then the basement reflection is at 175-msec two-way travel time in the VSP record, and should occur at 190- to 200-msec two-way travel time in the surface seismic section.

ACKNOWLEDGMENTS

We thank Hank Mullins for the opportunity and funding to participate in the Finger Lakes project. Comments from the anonymous reviewers and from Hank Mullins allowed us to improve and clarify the manuscript, and discussions with Mullins and with Rob Wellner helped in the development of the interpretation. Lee Leonard did some of the drafting, Kerry Swanson produced some of the prints from the figures, and Rob Wellner provided the surface seismic section and core results. The downhole logging data were plotted and analyzed using the Viewlog program; Dirk Kassenaar supplied regular Viewlog updates.

APPENDIX

The electrical properties across the bedding are akin to a set of resistors in series (e.g., Maillet, 1947; Nobes, 1984), so that the resistivity *perpendicular* to the bedding, ρ_v , is linearly related to the thicknesses and resistivities of the individual layers:

TABLE 1. TABULATION OF THE VSP AND SURFACE SEISMIC REFLECTOR TWO-WAY TRAVEL TIMES, WITH ESTIMATED DEPTHS, AND INTERPRETED ORIGIN AND NATURE OF THE REFLECTORS

Reflector	Origin and Nature of Event	Estimated Depth (m)	Two-way Travel Times VSP (msec)	Surface Seismic (msec)
A	Boundary between subunits 4f (silt) and 4e (sand/silty sand)	34	45	45 to 55
B	Top of sand in subunit 4e	39	55	55? to 65?
Unlabeled	Weak reflector in subunit 4d	49	65	60? to 65?
C	Boundary between subunits 4d (silt/clayey silt) and 4c (silty sand)	53	70	70 to 80
D	Boundary between subunits 4c (silty sand) and 4b (clay/clayey silt)	58	80	80 to 85
E	Boundary between subunits 4b (clay/clayey silt) and 4a (bedded silts)	64	85	90? to 95?
F	Bed within subunit 4a (bedded silts)	71	95	95 to 105
G	Boundary between subunits 4a (bedded silts) and 2b (sand/silty sand)	83	105	110 to 120
H	Sand bed within subunit 2b	89	115	120 to 130
I	Bed within subunit 2a (bedded silts)	101	125	135 to 140
J	Bed within subunit 2a (bedded silts)	111	135	155?
K	Boundary between subunit 2a (bedded silts) and unit 1 (Valley Heads Moraine)	122	145	165
L	Within unit 1, nature unknown	~130	160	175 to 180
M	Within unit 1, nature unknown	~138	170	185
N	Boundary between unit 1 (Valley Heads Moraine) and basement	~142	175	Not recorded but expected at 200 msec
P	Within basement, nature unknown	~155	185	Not recorded

$$\rho_v \Sigma_i t_i = \Sigma_i \rho_i t_i \quad (A1)$$

where ρ_i is the resistivity of layer i, and t_i is the thickness of layer i. Similarly, the electrical properties along the bedding are like those of a set of conductors in parallel, so that the conductivity *parallel* to the bedding, σ_h, is linearly related to the thicknesses and conductivities of the individual layers:

$$\sigma_h \Sigma_i t_i = \Sigma_i \sigma_i t_i \quad (A2)$$

where σ_i is the conductivity of layer i. We may then determine the perpendicular conductivity, σ_v, and the parallel resistivity, ρ_h, as the inverses of ρ_v and σ_h, respectively. The difference between the electrical properties in the directions parallel and perpendicular to bedding gives rise to an *anisotropy* which can be characterized by an anisotropy parameter, f, which is defined as (Maillet, 1947; Edwards et al., 1984):

$$f^2 = \sigma_h \rho_v = \rho_v / \rho_h \quad (A3)$$

The mean formation electrical resistivity, ρ, and conductivity, σ, are found to be:

$$\rho^2 = \rho_v \rho_h \quad \text{and} \quad \sigma^2 = \sigma_v \sigma_h \quad (A4)$$

REFERENCES CITED

Edwards, R. N., Nobes, D. C., and Gomez-Trevino, E., 1984, Offshore electrical exploration of sedimentary basins: The effects of anisotropy in horizontally isotropic, layered media: Geophysics, v. 49, p. 566–576.

Ellis, D. V., 1987, Well logging for earth scientists: New York, Elsevier Science Publishing, 532 p.

Hatton, L., Worthington, M. H., and Makin, J., 1986, Seismic data processing: theory and practice: Oxford, Blackwell Scientific Publishing, 177 p.

Jackson, P. D., Taylor Smith, D., and Stanford, P. N., 1978, Resistivity-porosity-particle shape relationships for marine sands: Geophysics, v. 43, p. 1250–1268.

Kassenaar, J. D. C., 1989, Automated classification of geophysical well logs [M.Sc. thesis]: Waterloo, Ontario, Canada, University of Waterloo, 120 p.

Kassenaar, J. D. C., 1991, An application of principal component analysis to borehole geophysical data: Proceedings, Fourth International Symposium on Borehole Geophysics for Minerals, Geotechnical and Groundwater Applications: Toronto, Ontario, Canada, p. 211–218.

Labo, J., 1986, A practical introduction to borehole geophysics: Tulsa, Oklahoma, Society of Exploration Geophysicists, 330 p.

Maillet, R., 1947, The fundamental equation of electrical prospecting: Geophysics, v. 12, p. 529–556.

McNeill, J. D., 1990, Use of electromagnetic methods for ground water studies, *in* Ward, S. H., ed., Geotechnical and environmental geophysics. Vol. I, Review and tutorial: Tulsa, Oklahoma, Society of Exploration Geophysicists, p. 191–218.

Mullins, H. T., and Hinchey, E. J., 1989, Erosion and infill of New York Finger Lakes: Implications for Laurentide ice sheet deglaciation: Geology, v. 17, p. 621–625.

Mullins, H. T., Wellner, R. W., Petruccione, J. L., Hinchey, E. J., and Wanzer, S., 1991, Subsurface geology of the Finger Lakes region, *in* Ebert, J. R., ed., Field Trip Guidebook, New York State Geological Association, 63rd Annual Meeting, Oneonta, State University of New York, p. 1–54.

Nobes, D. C., 1984, The magnetometric off-shore electrical sounding (MOSES) method and its application in a survey of upper Jervis Inlet, British Columbia [Ph.D. thesis]: Toronto, Ontario, Canada, University of Toronto, 206 p.

Telford, W. M., Geldart, L. P., and Sheriff, R. E., 1990, Applied geophysics (second edition): Cambridge, Cambridge University Press, 770 p.

Manuscript Accepted by the Society January 16, 1996

Printed in U.S.A.

Geological Society of America
Special Paper 311
1996

Late Pleistocene–Holocene lake-level fluctuations and paleoclimates at Canandaigua Lake, New York

Robert W. Wellner* and Thomas R. Dwyer*
Department of Earth Sciences, Heroy Geology Laboratory, Syracuse University, Syracuse, New York 13244

ABSTRACT

A series of four cores up to 13 m long are used to document the stratigraphic framework beneath the dry lake valley south of Canandaigua Lake, New York. Radiocarbon ages indicate a continuous, late-glacial and postglacial record that spans the last 13,650 ± 210 B.P. Deposition rates varied between 0.27 and 2.44 m/1,000 yr. Paleomagnetic and palynologic data support the radiocarbon data and allow linear interpolation between ages.

The first-order stratigraphy records the migration (transgression and regression) of open lacustrine (marl), freshwater marsh (peat), and floodplain (unfossiliferous clay) environments in response to lake-level fluctuations, which may have been climatically driven. Documented lake-level fluctuations occurred on a millennium-scale, with relative highstands centered at ~13,500, 12,600, 10,400, and 9,400 B.P. The maximum relative highstand occurred at ~10,400 B.P., when a shallow open lacustrine facies extended as much as 4.5 km south of modern-day Canandaigua Lake. Since 9,400 B.P., the relative lake level appears to have regressed steadily to its present-day position.

Pollen stratigraphy from Canandaigua Valley shows marked postglacial vegetational changes that support the interpretation of climatically driven lake-level fluctuations. An abrupt decline of both deciduous hardwood and boreal taxa in the interval 11,300 to 10,200 B.P. appears to record a brief cooling event that is contemporaneous with the Killarney/Younger Dryas events. The maximum lake-level highstand observed in this study also occurs during this interval, suggesting that wetter conditions prevailed during this time. By the start of the Holocene (~10,000 B.P.), a pine-dominated forest had become established indicating a general trend toward drier and likely warmer climatic conditions. This trend peaked between 7,300 and 2,300 B.P., when an oak-dominated, deciduous forest became established. Pollen percentages do not change significantly after 1,400 B.P., which implies that climatic conditions became similar to the modern at this time.

INTRODUCTION

To better understand past, and potentially predict future, climate change, high-resolution stratigraphic records are needed of glacial to postglacial periods. Many early attempts to relate late glacial climatic events in northeastern North America (Deevey, 1939; Leopold, 1956; Ogden, 1967) to the well-documented paleoclimates and chronostratigraphy of Europe and the British Isles are not considered convincing. The evidence presented from isolated sites is often unequivocal, reflecting local environmental events, rather than widespread climatic change (Mott

*Present addresses: Wellner, Exxon Production Research Company, P.O. Box 2189, Houston, Texas 77252-2189; Dwyer, Blasland, Bouck and Lee, Inc., 6723 Towpath Road, Syracuse, New York 13214.

Wellner, R. W., and Dwyer, T. R., 1996, Late Pleistocene–Holocene lake-level fluctuations and paleoclimates at Canandaigua Lake, New York, *in* Mullins, H. T., and Eyles, N., eds., Subsurface Geologic Investigations of New York Finger Lakes: Implications for Late Quaternary Deglaciation and Environmental Change: Boulder, Colorado, Geological Society of America Special Paper 311.

et al., 1986). However, recent palynologic (Davis et al., 1980; Davis, 1983; Shane, 1987; Peteet et al., 1990; Baker et al., 1992) and lake-level studies (Hansel and Mickelson, 1988; Tinkler et al., 1992) from the central and eastern United States argue for regionally extensive climatic fluctuations during the late Wisconsin and Holocene that are similar to those observed in Europe (Lowe et al., 1988; Watts, 1980; Ammann and Lotter, 1989), Greenland (Dansgaard et al., 1982, 1989), eastern Canada (Mott et al., 1986; Rawlence and Senior, 1988; Walker et al., 1991); North Atlantic Ocean (Ruddiman and McIntyre, 1981; Broecker et al., 1990; Wright, 1989), and Pacific Ocean (Mathewes et al., 1993).

Much emphasis has been placed on the period of rapid deglacial warming between 13,000 and 11,000 B.P. (European Bölling and Allerod chronozones), immediately following Termination 1a in the North Atlantic oxygen isotope record (Duplessy et al., 1981). An abrupt return to colder conditions, termed the Younger Dryas event, between ~11,000 and 10,000 B.P. has also been of considerable interest mainly because it began and ended quickly with large climatic shifts that signaled the return to near full-glacial conditions within several human lifespans (Flohn, 1986). Levesque et al. (1993) recently presented evidence from the Canadian Maritime provinces for a short-lived cold period between 11,160 and 10,910 B.P., termed the Killarney Oscillation, which predates the Younger Dryas. Although the cause of the Killarney and Younger Dryas events remains uncertain, these events are generally attributed to oceanographic circulation changes (Ruddiman, 1987) associated with either meltwater flow to the North Atlantic Ocean (Broecker et al., 1988) or to oscillations in oceanic salt budgets (Broecker et al., 1990). The rapid input of glacial meltwater to the northern Atlantic Ocean may have reduced the salinity of the ocean water, which in turn may have caused the ocean's thermohaline circulation to slow or possibly stop, thereby greatly reducing the flux of heat to the northern latitudes between 11,000 and 10,000 B.P. (Broecker and Denton, 1989).

The return to drier and/or warmer conditions at the end of the Younger Dryas event (oxygen isotope Termination 1b) signals the beginning of the Holocene (~10,000 B.P.) when there was rapid vegetational and environmental change (Street-Perrott and Harrison, 1984; Webb et al., 1987; Shane, 1987; Ammann and Lotter, 1989; Overpeck, 1991).

Conceptual models of paleoclimatic change are generally constructed using quantitative pollen and lake-level data that are supported by robust chronologies (e.g., radiocarbon ages, paleomagnetic data) (COHMAP Members, 1988). Unfortunately, many of the late-glacial and postglacial pollen (McCulloch, 1939; Deevey, 1943; Cox, 1959; Cox and Lewis, 1965; Miller, 1973; Spear and Miller, 1976; O'Rourke, 1976; Miller, 1988), paleomagnetic (King et al., 1983; Brennan et al., 1984; Carmichael et al., 1990; Ridge et al., 1990), and lake-level studies (Muller and Prest, 1985; Anderson and Lewis, 1985; Flint et al., 1988; Weninger and McAndrews, 1989; Dalrymple and Carey, 1990) from New York State and Lake Ontario have suffered from either the lack of a complete late-glacial to postglacial record and/or poor chronologic control.

In an effort to obtain a complete late-glacial through Holocene climate record from the Finger Lakes region of New York State, a series of drillcores (118 and 12 m) and handcores (10 m) were collected from the dry lake valley south of Canandaigua Lake (Fig. 1). The objectives of this study are to: (1) establish a high-resolution record of paleoenvironmental change within the Finger Lakes region during the past 14 ka, (2) evaluate the subsurface stratigraphy in terms of relative lake-level changes, and (3) tie the Canandaigua Lake data to the northeastern North America paleoclimatic data base.

SETTING

The 11 lakes that comprise the Finger Lakes of central New York State (Fig. 1) occur along the northern margin of the glaciated Appalachian Plateau and are cut principally into undeformed, but well-jointed, Devonian sedimentary rocks (largely shale) that dip gently to the south-southwest (Coates, 1974). Canandaigua Lake is an oligotrophic lake 25 km in length, with a maximum width of 2.5 km (Fig. 1). This lake is 210 m above mean sea level and has a maximum water depth of 84 m (the modern lake level is controlled by several artificial dams that control the northward outflow of water through Canandaigua Outlet) (Eaton and Kardos, 1978).

To the south of Canandaigua Lake, a relatively flat, dry lake valley extends some 8 km before the Valley Heads Moraine rises more than 200 m above the valley floor (Fig. 1). Based on morphostratigraphic correlation (Fullerton,1986), truncation by the Valley Heads Moraine of radiocarbon-dated moraines in eastern New York (Krall, 1977), and radiocarbon ages from outwash associated with the Valley Heads Moraine (Muller and Calkin, 1993), deposition of the moraine is interpreted to have occurred between 14,800 and 14,400 B.P.

The modern climate of the Finger Lakes region is marked by warm summers and long, cold winters. Annual precipitation ranges between 66 and 97 cm, with an annual mean of 84.1 cm; average annual snow fall is ~158 cm (Eaton and Kardos, 1978).

According to Eaton and Kardos (1978), the forest cover, prior to deforestation by settlers (ca. 1820 A.D.), consisted of hardwoods, pine, and hemlock. The present-day upland vegetation consists mainly of oak (*Quercus*) and hickory (*Carya*), although more shaded areas, such as the numerous gullies (Fig. 1), contain hemlock (*Tsuga*), birch (*Betula*), beech (*Fagus*), and maple (*Acer*). Basswood (*Tilia americana*), black walnut (*Julgans nigra*), and white ash (*Fraxinus americana*) are also common, while red cedar (*Juniperus virginianus*) and white pine (*Pinus strobus*) are less common. Sycamore (*Platanus occidentalis*), willows (*Salix*) of various species, and cottonwood (*Populus deltoides*) are common trees along the edge of the lake. In High Tor Wetlands at the southern end of the lake (Fig. 1), red maple (*Acer rebrum*) and black ash (*Fraxinus nigra*) dominate, along with buttonbush (*Cephalanthus occidentalis*) and cattail (*Typha latifolia*). Dead American elm (*Ulmus americana*) trunks are common.

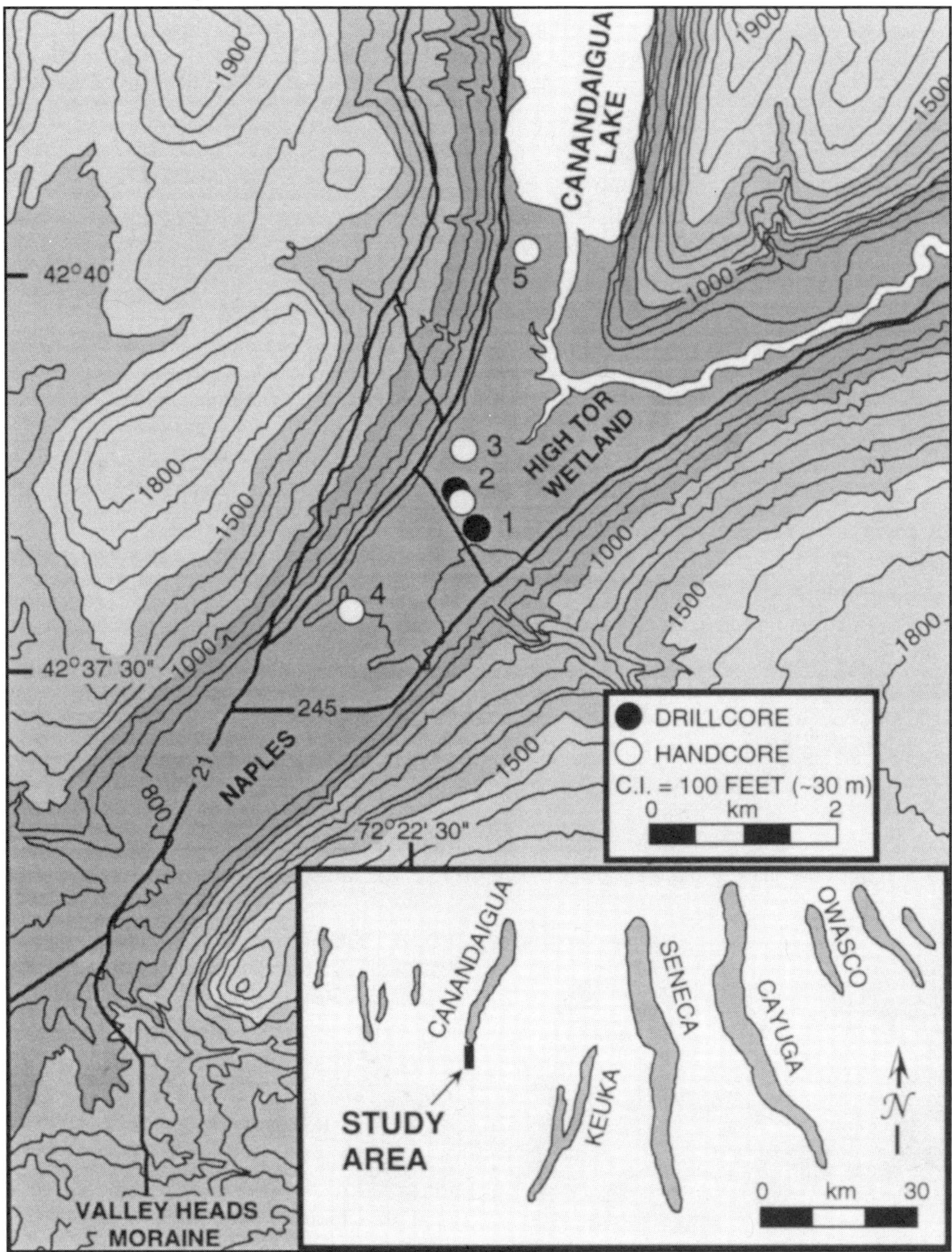

Figure 1. Location map of southern portion of Canandaigua Lake and adjacent dry valley illustrating location of drillcores (cores 1 and 2) and handcores (cores 3 through 5). Note position of modern High Tor Wetland (HTW) and the Valley Heads Moraine (VHM) relative to the core sites. Topographic contour interval is 100 ft (~30 m).

METHODS

During the summer of 1990, two drillcores (5 cm in diameter) were collected in the dry lake valley 3 km south of Canandaigua Lake using a truck-mounted platform and rotary core and wire-line techniques. Site 1 extended 118 m into the subsurface; site 2, located 168 m to the west, extended only 13 m into the subsurface, but recovered a nearly complete upper section (Fig. 1). Due to the higher core recovery (>90%) at drillsite 2 (Fig. 1), more extensive analysis was conducted on this core. In addition, three handcores (<10 m) were collected using a spilt-spoon soil sampler (2 cm in diameter) during the fall of 1992. This was done to laterally extend the subsurface stratigraphy defined at drillsite 2 (Fig. 1).

The drillcores were split, described, photographed, and radiographed. The upper 12 m of core 2 (Fig. 1) was sampled every 10 cm and analyzed for total organic carbon (TOC) using a LECO induction furnace and total carbonate content by acidification. Handcores (Fig. 1) were analyzed for total organic and carbonate content by loss on ignition (LOI) following the procedures of Dean (1974).

Paleomagnetic measurements were carried out on 1-cm³ samples from the upper 12 m of the split rotary drillcores. Natural remanant magnetization (NRM) directions and intensities were measured using a Schonstedt fluxgate spinner magnetometer (model SSM-2). Samples were magnetically cleaned up to 1,000 Oe using a Sapphire Instruments alternating field demagnetizer (model SI-4).

Sediment samples for pollen analysis were collected at ~20-cm intervals from core 2 between depths of 11.7 and 2.5 m. Because of the poor pollen preservation in the upper 2.5 m of oxidized clay in core 2, samples were obtained from peat in handcore 5 between depths of 2.5 and 0.05 m. Samples were prepared for pollen analysis using the procedure of Faegri and Iversen (1975). A minimum of 300 known grains were identified and counted for each sampled interval. Typical pollen grains observed are illustrated in Figure 2.

Nine in-situ clayey to sandy peat layers were sampled from core 2 and sent to the University of Texas, Austin, Radiocarbon Lab for conventional radiocarbon dating (Table 1). Single samples were collected from handcore 4 (2.05 m) and handcore 5 (7.45 m) and sent to Geochron Laboratories for accelerator mass spectrometer (AMS) dating at its New Zealand facilities (Table 1). All radiocarbon ages are based on the Libby half-life of 5,570 yr. Samples were pretreated with acid to remove carbonate, and final ages have been corrected for sample $\delta^{13}C$ fractionation.

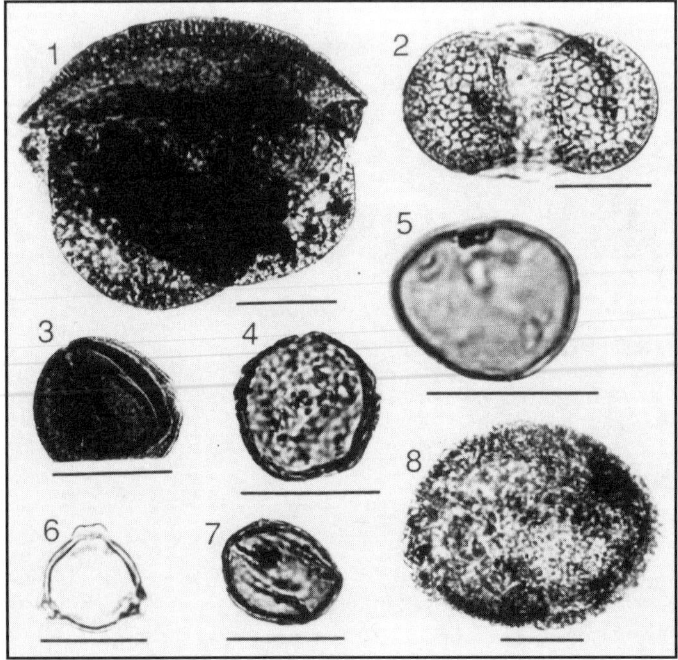

Figure 2. Light micrographs of typical pollen grains observed in Canandaigua valley cores. 1 = fir (*Abies*); 2 = pine (*Pinus*); 3 = beech (*Fagus*); 4 = elm (*Ulmus*); 5 = grass (*Gramineae*); 6 = birch (*Betula*); 7 = oak (*Quercus*); 8 = hemlock (*Tsuga*). Scale bars = 25 μ.

TABLE 1.CONVENTIONAL AND ACCELERATOR MASS SPECTROMETRY ^{14}C DATES OF CANANDAIGUA VALLEY CORE SAMPLES

Core	Depth (m)	Radiocarbon Age (B.P.)	$\delta^{13}C$ (%)	Laboratory Number
2	3.78	4,840 ± 280	-28.1	UTA7254
2	6.68	9,000 ± 500	-30.0	UTA7256
2	8.18	9,130 ± 220	-30.3	UTA7257
2	8.68	9,730 ± 440	-28.8	UTA7258
2	9.44	10,180 ± 240	-29.9	UTA7259
2	10.15	12,930 ± 570	-30.0	UTA7260
2	11.18	13,250 ± 1,180	-30.2	UTA7261
1	12.0	13,650 ± 210	-28.5	UTA7253
1	5.6	5,060 ± 530	-27.9	UTA7252
5	2.15	2,477 ± 72	-27.8	GX18539-AMS
4	7.45	10,314 ± 108	-29.0	GX18358-AMS

RESULTS

Lithostratigraphy

Analysis of the Canandaigua Valley subsurface stratigraphy reveals several distinct depositional facies that can be related to modern depositional environments within the basin. The observed facies include: (1) unfossiliferous, oxidized, gray clay to sandy clay with low total organic matter (<5%) and carbonate (<5%) values (Fig. 3); (2) loose to densely compacted woody peat with high total organic matter (>85%) and low carbonate (<10%) values (Fig. 3); (3) marl that has low total organic matter (<10%) and high carbonate (up to 37%) values (Fig. 3); and (4) a transitional facies between facies 2 and 3 that is characterized by isolated gastropods and both intermediate total organic matter (5 to 10%), and carbonate (3 to 6%) values (Fig. 3).

The lack of freshwater gastropods, the scarcity and poor preservation of pollen grains, and evidence of oxidation suggests that facies 1 was deposited in a subaerial environment, most likely as overbank (floodplain) deposits. Facies 2 is interpreted as freshwater marsh deposits similar to the modern marsh in High Tor Wetlands (Fig. 1). Facies 3 has abundant freshwater gastropods that Dwyer and Darling (1993) identified as *Valvata tricaranata*, *V. sincera*, *Physid gryina*, *Hydrobiidae* sp., and *Planorbidae* sp.). The freshwater pelecypod (*Musculium* sp.) is also been present. The *Valvata* and *Hydrobiidae* gastropods are gill-breathers and respire subaqueously, whereas the *Physid* and *Planordid genera* are lung-breathers and need emergent vegetation (Good, 1987). The observed fossil assemblage is indicative of a shallow, low-energy lacustrine environment with emergent vegetation (Good, 1987). Facies 3 is broadly interpreted as an open lacustrine deposit. Facies 4 consists of peaty marl with isolated gastropods and is interpreted as a shallow lacustrine-marsh interface deposit.

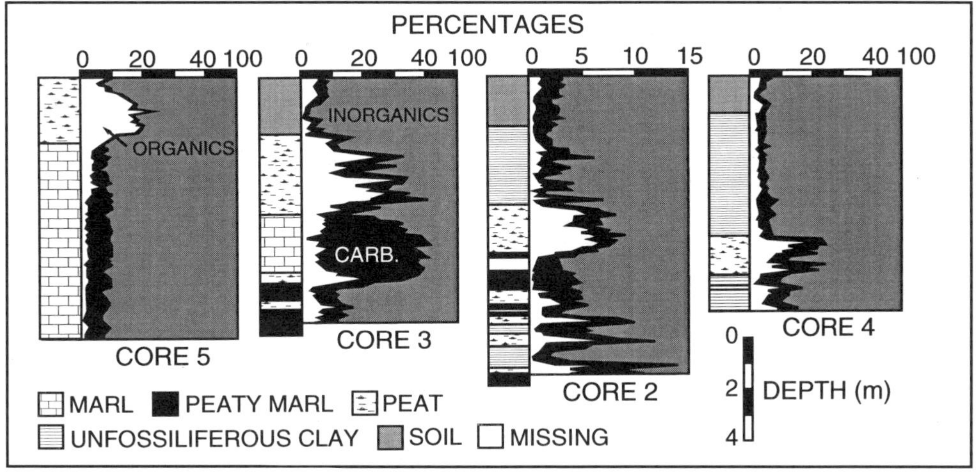

Figure 3. Percentage of organic material, total carbonate, and inorganic content of the Canandaigua Valley drill- and handcores. Note that core 2 was analyzed using a LECO induction furnace and total carbonate was measured by acidification. Cores 3 through 5 were analyzed using the procedures of Dean (1974).

Chronostratigraphy

Radiocarbon ages. The radiocarbon ages obtained from in situ peat layers in core 2 (Table 1) indicate nonlinear accumulation rates that ranged from 0.27 to 2.44 m/1000 yr over the last 13,650 ± 210 B.P. (Fig. 4).

The $\delta^{13}C$ values from the core 2 peat layers (Table 1) range between –27.9 and –30.3‰, which is typical for emergent aquatic plants with C_3 photosynthetic cycles (Aravena et al., 1992). Emergent aquatic plants are less susceptible to potential hard water reservoir effects, and thus yield more reliable radiocarbon ages than lacustrine-precipitated $CaCO_3$ or submerged vegetation (Aravena et al., 1992).

Based on an AMS radiocarbon age of 2,477 ± 72 B.P. from 2.15-m subsurface depth in core 5 (Table 1; Fig. 4) and assuming a zero age for the core top, an accumulation rate of 0.82 m/1000 yr is calculated. This is similar to the accumulation rate of 0.85 m/1000 yr determined for core 2 (Fig. 1) between subsurface depths of 0 and 4.42 m (Fig. 4).

A paleomagnetic investigation of the core 2 sediments was undertaken to determine the secular variation of paleoinclination. Such a paleoinclination record from Canandaigua Valley may be used as an independent check of the radiocarbon chronology, via comparison to similar, radiocarbon-dated records from the Great Lakes region (e.g., Creer and Tucholka, 1983). Secular variation of paleodeclination could not be accurately determined because the Canandaigua Valley drillcores were not oriented azimuthally.

The magnetic stability of the sediment in drillcore 2 (Fig. 1) was assessed by subjecting random samples to alternating field

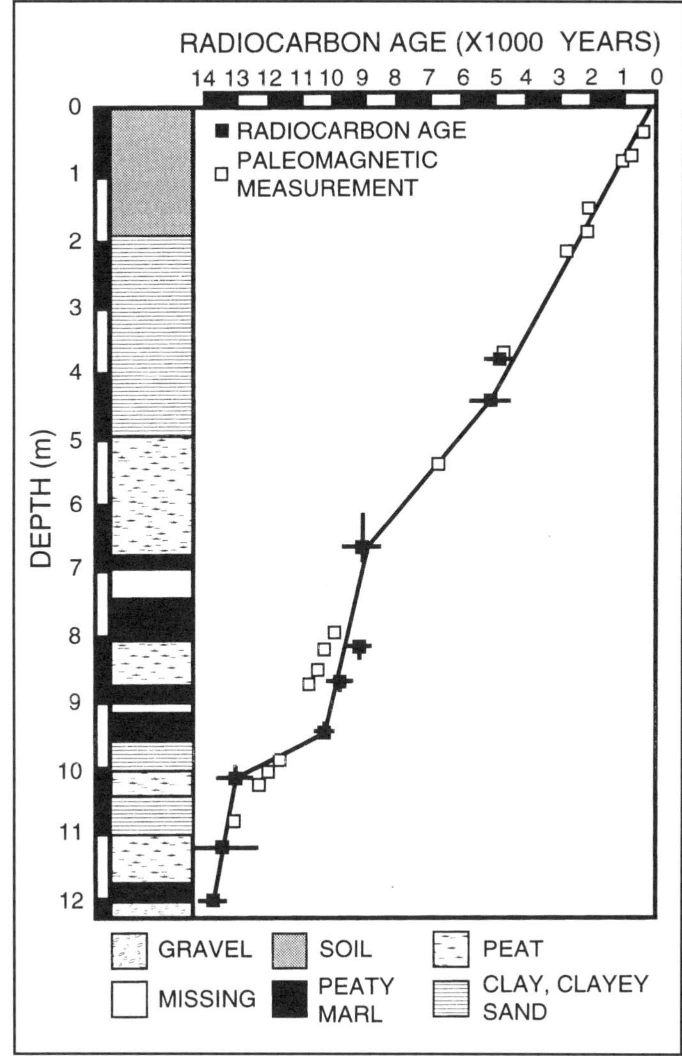

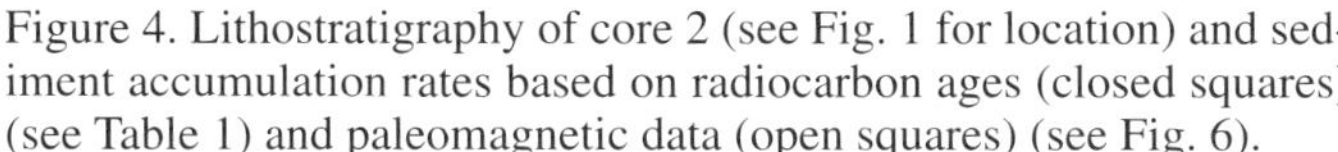

Figure 4. Lithostratigraphy of core 2 (see Fig. 1 for location) and sediment accumulation rates based on radiocarbon ages (closed squares) (see Table 1) and paleomagnetic data (open squares) (see Fig. 6).

(AF) demagnetization. Organic-poor clays at 2.70 m and organic-rich clays at 9.61 m exhibit a soft secondary magnetization or viscous remanent magnetization (VRM), that is removed by 100-Oe peak fields (Fig. 5). An isothermal remanent magnetization (IRM) plot of the sample from 2.70 m (Fig. 5) reveals that saturation magnetization occurs at ~±1,000 Oe and remanence coercivity occurs at ~–500 Oe, indicating that magnetite is the primary magnetic carrier in this sample (Fig. 5). Total magnetization values of the 64 samples analyzed ranged from 2.82×10^{-5} to 1.10×10^{-7} emu/cc.

The natural remanent magnetization (NRM) of the Canandaigua Valley core sediments is believed to be reliable recorder of postglacial secular variations. A record of secular variation of paleoinclination has been constructed for Canandaigua Valley (Fig. 6). This record is similar to the composite, stacked paleoinclination curve from the Great Lakes region (Creer and Tucholka, 1982) that has been independently dated using radiocarbon ages. By correlating similar magnetic excursions (Fig. 6), the Canandaigua Valley depth scale has been independently converted to time. Based on these correlations, the paleoinclination data from Canandaigua Valley has been used to determine a sediment accumulation rate of 0.84 m/1,000 yr for the shallow stratigraphy (subsurface depth, <3.5 m) in core 2 (Fig. 3) where no datable organic material occurs. The paleomagnetic data also indicate a relatively constant accumulation rate between the radiocarbon-dated peat layers.

Pollen

The first-order, postglacial vegetational changes at Canandaigua Lake can be inferred from the relative pollen sum diagram (Fig. 7). In order to facilitate comparisons to nearby regions, the pollen diagram has been split into zones T, A, C-1, C-2, and C-3, in a fashion similar to that used by Spear and Miller (1976) for sites in western New York State.

Zone T (>~13,100 B.P.). The presence of of boreal taxa, such as spruce (*Picea*), pine, and fir, in conjunction with relatively high pollen percentages of grass and hemlock, indicate that this basin was a primarily open, but not treeless, park tundra prior to 13,100 B.P. (Fig. 7).

Zone A (~13,100–9,940 B.P.). The boundary between pollen zones T and A is defined by the increase of boreal and hardwood taxa and decreased grass values at ~13,100 B.P. Spear and Miller (1976) reported an age of 12,565 ± 115 B.P. and Peteet et al. (1990) reported an age of 12,840 ± 110 B.P. for Belmont Bog, New York, and Alpine Lake, New Jersey, respec-

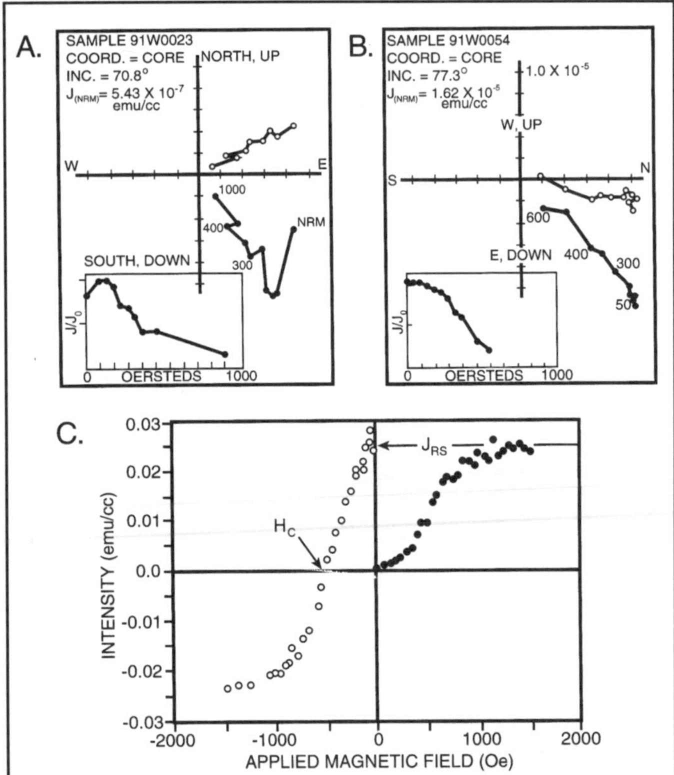

Figure 5. A and B, Vector difference diagrams demonstrating the magnetic stability of the Canandaigua Valley sediments. C, Induced remanent magnetization (IRM) curve. Saturation magnetization (J_{rs}) occurs at ~1000 Oe and demagnetization coercivity (H_c) occurs at ~500 Oe, which indicate that the mineral magnetite is the primary magnetic carrier.

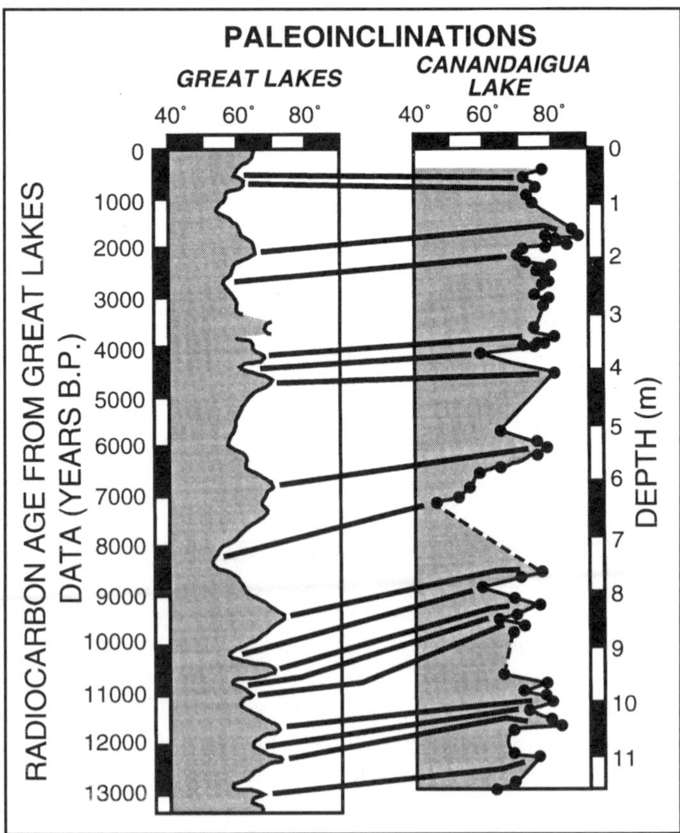

Figure 6. Secular variation curves (<14,000 ^{14}C B.P.) of magnetic inclination for the Great Lakes (Creer and Tucholka, 1982) and Canandaigua Valley. Canandaigua Valley depth scale is converted to age via correlation of excursions with independently dated Great Lakes composite curve. Magnetic data are used to determine accumulation rates for the upper ~3 m of core 2 and between radiocarbon dates. (See Fig. 4.)

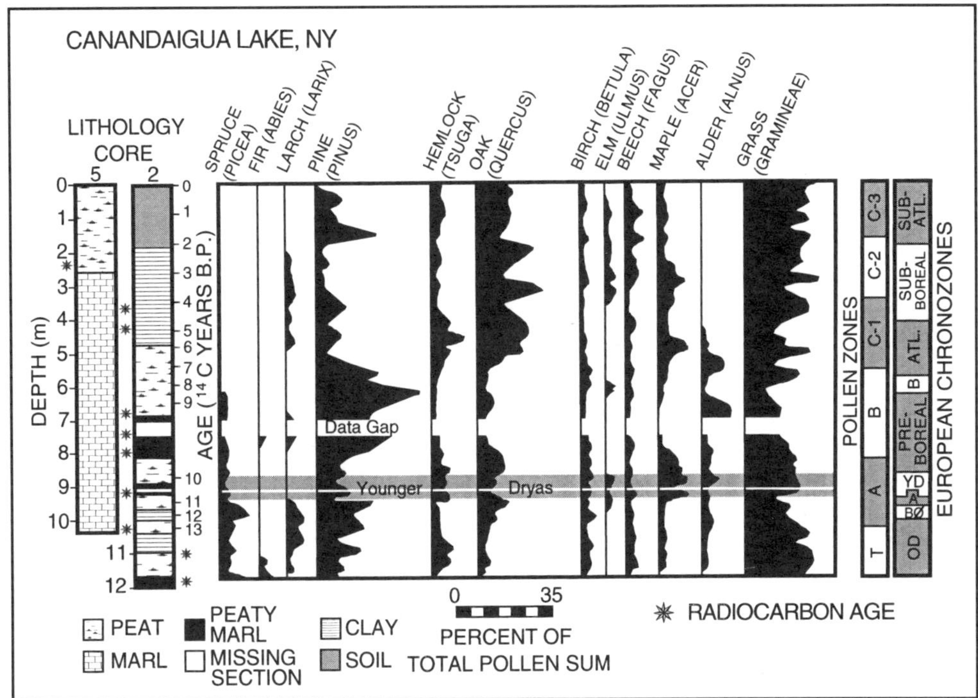

Figure 7. Relative pollen frequency diagram. Pollen grains were counted approximately every 20 cm in core 2 from a subsurface depth of 11.75 to 2.5 m, and the upper 2.5 m of core 5. Correlation among cores is achieved via radiocarbon ages (10^3 yr).

tively, based on similar criteria. This zone is characterized by the high spruce (maximum, 19%) and larch (*Larix*) (maximum, 12%) pollen values, as well as moderately high pine (>20%) (Fig. 7). Both boreal (e.g., pine and spruce) and thermophilous taxa (e.g., oak) exhibit a synchronous and rapid decline by 11,300 B.P. (Fig. 7). Pine and oak pollen percentages remain low (<~30% of total pollen sum) until ~9,900 B.P (Fig. 7). Hemlock and grass, and to lesser extent alder (*Alnus*), pollen percentages increase during this interval (Fig. 7).

The gradual increase of boreal and deciduous hardwood taxa, with the exception of maple, prior to ~11,000 B.P., indicates that a mixed forest had developed, most likely in response to a relatively cool, but warming, humid climate (e.g., Peteet et al., 1993). The synchronous decline of boreal and thermophilous taxa and expansion of grass and hemlock at ~11,000 B.P. appears to signal an abrupt vegetational reversal that likely occurred in response to colder and possibly wetter conditions at this time (Peteet et al., 1990). Cold and wet conditions appear to have prevailed until ~9,900 B.P., after which the rapid expansion of pine and thermophilous taxa, such as oak, indicate a renewed trend toward a drier, and likely warmer, climate at the start of the Holocene (Webb et al., 1993).

Pollen zone B (9,840–7,500 B.P.). The boundary between pollen zones A and B is at 9,840 B.P. based on the abrupt rise of pine values. The dominance of pine in the Finger Lakes region after ~9,840 B.P. (Fig. 7) is similar to that observed from other areas in northeastern North America (e.g., Roger Lake, Connecticut [Davis, 1969]; West Thornton, New Hampshire [Davis et al., 1980]; Brandreth Bog, New York [Overpeck, 1985]; Lake Ontario [Anderson and Lewis, 1985]; Kitchener, Ontario [Fritz et al., 1987]; Heart Lake, New York [Whitehead et al., 1989]; and Alpine Swamp and Allamuchy Pond, New Jersey [Peteet et al., 1990; 1996]), but is considerably younger than the age of 10,945 ± 65 B.P. reported by Spear and Miller (1976) for the rapid expansion of pine at Belmont Bog, New York.

The abrupt shift from a spruce-pine–dominated woodland at the end of pollen zone A to a pine-dominated woodland by ~9,650 B.P. (Fig. 7) implies a drying, and likely a warming, climatic trend (Webb et al., 1987). The burial of highstand lake-level peaty marls by peat at core site 2 (Fig. 3) similarly argues for a drier climate, and lower lake levels, during this interval.

Pollen zone C, subzone C-1 (7,500–4,250 B.P.). The boundary between pollen zone B and subzone C-1 is defined by the decline of pine to <30% of the total pollen sum, and a concurrent increase of oak (maximum, 38%), maple (maximum, 24%), and hemlock (maximum, 19%) values (Fig. 7). The age of this boundary at Canandaigua Valley is similar to that reported for Belmont Bog, New York (~7,215 B.P., based on linear interpolation between radiocarbon ages; Spear and Miller, 1976), but is considerably younger than the ~9,000 B.P. reported by Peetet et al. (1990) from Alpine Swamp, New Jersey.

The abrupt rise of thermophilous taxa and synchronous decline of pine in pollen zone C-1 indicates that the drying and warming trend observed in pollen zone B continued in zone C-1 (Webb et al., 1987).

The decline of hemlock at ~4,800 B.P. at Canandaigua Valley correlates well with a similar period of rapid decline at ~4,700 B.P. observed throughout northeastern North America (Davis, 1981; Allison et al., 1986). The demise of hemlock at this time is thought to be the the result of a catastrophic pathogen outbreak similar to the Chestnut blight and Dutch-elm epidemics (Allison et al., 1986). Because the decline was synchronous throughout the range of hemlock, it can be used as a time-stratigraphic horizon. The similarity between the age of this event at Canandaigua Valley and similar, independently dated, events elsewhere, offers further evidence for the validity of the radiocarbon dates presented in this study.

Subzone C-2 (~4,250–2,100 B.P.). The decline of hemlock pollen values at ~4,250 B.P. defines the boundary between subzones C-1 and C-2 at Canandaigua Valley (Fig. 7). This subzone is characterized by decreasing pine and alder, increasing oak, elm, and maple, and the highest observed and most widely fluctuating oak values observed (Fig. 7). Larch, birch, and beech values also increased slightly at the beginning of this zone (Fig. 7). The increase of beech and oak implies that the warming trend observed in zone C-1 persisted, and likely intensified, into this zone (Webb et al., 1987).

Subzone C-3 (2100 B.P. through the present day). The boundary between subzones C-2 and C-3 is defined by the abrupt decline of oak to <25% of the total pollen sum; an increase in pine, particularly at 1,970 B.P., and a subtle increase—to greater than 10% of the total pollen sum—of hemlock. The increase of pine and hemlock at ~2,100 B.P. may imply that subzone C-3 was cooler and possibly wetter than subzone C-2 (Webb et al., 1987). Because pollen percentages do not change significantly after 1,400 B.P., the climate is interpreted to have become similar to the modern at Canandaigua Lake at this time.

DISCUSSION

The stratigraphic record at Canandaigua Lake (Fig. 3) may be the result of several local and/or regional factors, including: (1) the repetitive uplift and depression of the Canandaigua Lake outlet, following retreat of the Laurentide Ice Sheet, by differential isostatic adjustment and the northward propagation of the peripheral isostatic bulge; (2) episodic erosion of the outlet; and (3) climatically induced net variation of precipitation and evaporation.

Because of the smooth, largely unidirectional, response of the asthenosphere to ice unloading (Andrews, 1970; Walcott, 1972; Peltier and Andrews, 1976; Peltier, 1987), it is unlikely that the observed stratigraphy could be the result of differential isostatic rebound between the northern lake outlet and the southern inlet. In this scenario, relative lake highstands would occur when isostatic rebound was greater in the northern portion of the basin relative to the south, thus reducing water flow from the basin and raising the lake level. Episodic erosion of the northern outlet may have contributed to the relative lake-level falls, but could not produce lake-level rises. The pollen stratigraphy at Canandaigua Lake indicates several long-term, first-order vegetational changes that can be correlated to previously reported regional and global climatic variations (e.g., Younger Dryas event). Many of these vegetational changes are synchronous with lithostratigraphic changes observed in the Canandaigua Lake cores (Fig. 7). On the basis of these observations we propose that climatically driven lake-level fluctuations caused the migration (transgression and regression) of depositional environments, and the resulting stratigraphy observed beneath Canandaigua Valley (Fig. 8)

Four cycles of lake-level change, with frequencies on the order of 800 to 1,200 B.P., occurred during the last glacial-interglacial transition (Fig. 9). Relative lake-level highstands are centered at approximately 13,500, 12,600, 10,400, and 9,400 B.P. (Fig. 9). The maximum lake-level highstand at ~10,400 B.P. caused depositional environments to transgress nearly 4.5 km south of present-day Canandaigua Lake. After 9,400 B.P., the cores record the largely regressive backfilling of the basin (Figs. 8 and 9). Modern-day Canandaigua Lake appears to be at its most regressed position since 13,650 B.P.

After highstand proglacial lake levels dropped abruptly (see Mullins and Hinchey, 1989), the southern end of Canandaigua Lake was flooded in response to glacioisostatic rebound related to ice withdrawal from the basin (Tarr, 1904). As the Canandaigua Lake outlet underwent differential isostatic rebound, the amount of water leaving the basin decreased, and water levels began to rise in the south. Based on the radiocarbon dates from basal peats in cores 1 and 2, the initial flooding of the southern end occurred shortly before 13,650 ± 210 B.P. Continued deglacial isostatic rebound provided the accommodation space required to deposit the ~12-m-thick sequence of sediments south of Canandaigua Lake. This estimate for isostatic adjustment at Canandaigua Lake is similar to the ~14 m of isostatic adjustment proposed by Clark and Personage (1970) for similar latitudinal positions in western New York State.

The transition from park tundra to a spruce-pine–dominated boreal forest to a mixed hardwood-conifer forest during the late-glacial (~13,420 to 11,400 B.P.) indicates an overall warming trend through this period (Webb et al., 1987). These vegetational changes are equivalent to the Bölling and Allerod chronozones of Europe. The interpretation of an overall warming trend during this interval at Canandaigua Valley correlates well with a period of rapid deglacial warming observed in oxygen isotope records from the North Atlantic Ocean (e.g., Duplessy et al., 1981), Greenland ice cores (e.g., Oescheger et al., 1984), and European lake cores (e.g., Eicher, 1980; Lister, 1987), as well as with proxy climatic data based on vegetational shifts observed during the Bölling and Allerod chronozones on the European continent (Welten, 1982; Atkinson et al., 1987).

By 11,400 B.P., rising lake level (Fig. 9) resulted in the burial of lowstand mineral-rich sediments by peat (marsh) and ultimately peaty marl (shallow lacustrine environment) at core site 2 (Figs. 1 and 8). Based on an AMS radiocarbon date of 10,314 ± 108 B.P. from the base of a peat layer in core 4 (Table 1;

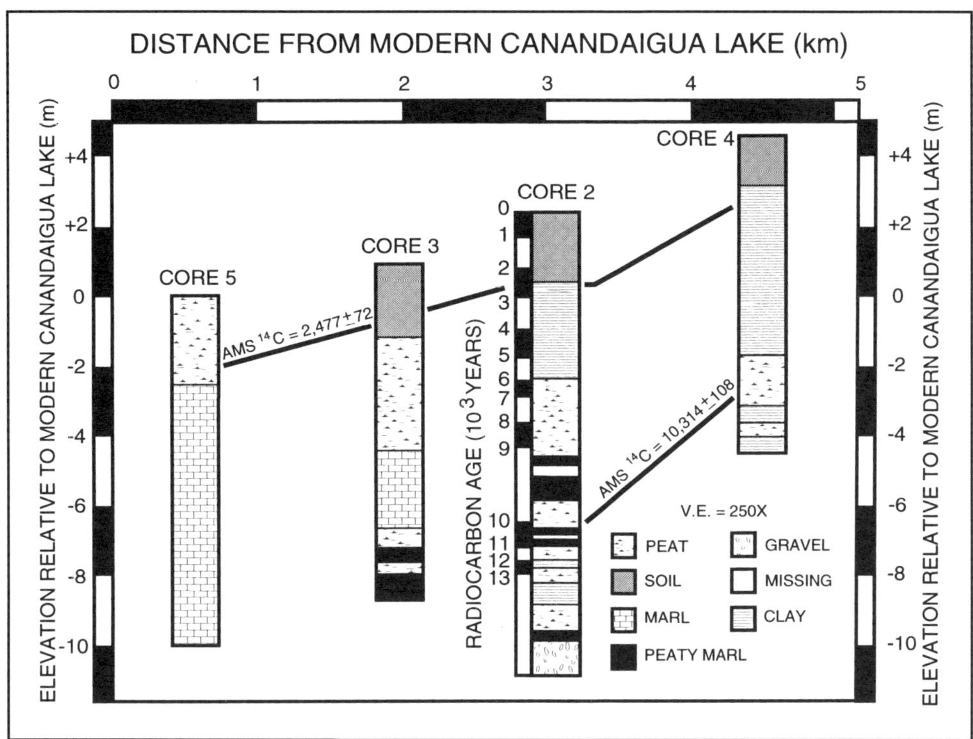

Figure 8. Stratigraphy beneath the dry valley south of Canandaigua Lake with cores plotted relative to the surface of modern Canandaigua Lake (210 m above sea level). Radiocarbon age scale is shown for core 2. Time lines based on radiocarbon age dates from cores 4 and 5 are also indicated.

Fig. 8), the paleolake shoreline is interpreted to have been some 3 to 4.5 km south of modern-day Canandaigua Lake (Figs. 1 and 9).

Near the beginning of this highstand interval (~11,300 B.P.), there is a sharp decline of boreal and thermophilous tree taxa and a synchronous increase of hemlock, and grass pollen values (Fig. 7). The abrupt lithologic and vegetational changes appear to signal a return to colder and possibly wetter conditions (Webb et al., 1987) in the Finger Lakes region at this time. The identification of rapid vegetational shifts at ~11,300 B.P. correlates well with the Killarney/Younger Dryas events and adds to the growing documentation of a regionally extensive climatic shift to colder conditions in northeastern North America at this time. By ~10,000 B.P., maple, oak, and pine values have increased at the expense of hemlock pollen (Fig. 7). This indicates a renewed trend toward drier, and likely warmer, climatic conditions at the start of the Holocene (Webb et al., 1987).

At ~10,000 B.P., lake levels began to drop, and as a result, shallow lacustrine, peaty marl is buried by marsh peat at core site 2 (Figs. 1 and 8). At the beginning of this lowstand interval, pine pollen percentages begin to rapidly increase, as climatic conditions became apparently drier (Webb et al., 1987) than during the previous interval in Canandaigua Valley (Fig 7). This proposed trend at the start of the Holocene correlates well with vegetational changes previously recognized throughout northeastern (e.g., Davis et al., 1983; Peteet et al., 1990) and central North America (e.g., Shane, 1987).

A relative lake-level highstand beginning at ~9,670 B.P. (Fig. 9) resulted in the burial of marsh peat by shallow lacustrine, peaty marl with freshwater gastropods at core site 2 (Figs. 1 and 8). During the initial portion of this relative lake highstand interval, pine values increase, while thermophilous taxa, such as oak and maple, decrease (Fig. 7). Unfortunately, the section of core spanning ~9,400 to 9,250 B.P. was not recovered (Fig. 7). However, pollen samples from either side of the missing interval indicate that pine values had decreased, while alder increased to its highest observed values. These vegetational and lithostratigraphic changes may indicate that the drying and warming trend observed at the start of the Holocene was interupted by a brief period of more wet climatic conditions.

After ~9,200 B.P., lake level is interpreted to have fallen past core site 2, based on the burial of previous lake highstand sediments by peat. The interpretation of drier climatic conditions through this interval is supported by the synchronous and rapid increase of pine pollen at this time (Fig. 7) (Webb et al., 1987).

The overall decline of conifer pollen values and the dominance of thermophilous taxa, such as oak, by ~6,450 B.P. indicate enhanced heating and drying of Canandaigua Valley through middle Holocene (Webb et al., 1987). After ~1,400 B.P., the pollen record appears to stabilize, suggesting that the climate became very similar to the modern climate at this time.

Collectively, the lithostratigraphic and vegetational changes during the Holocene at Canandaigua Valley record a drying and warming trend similar to that observed in the western Great Lakes region (Wright, 1968; Bernabo and Webb, 1977; Keen

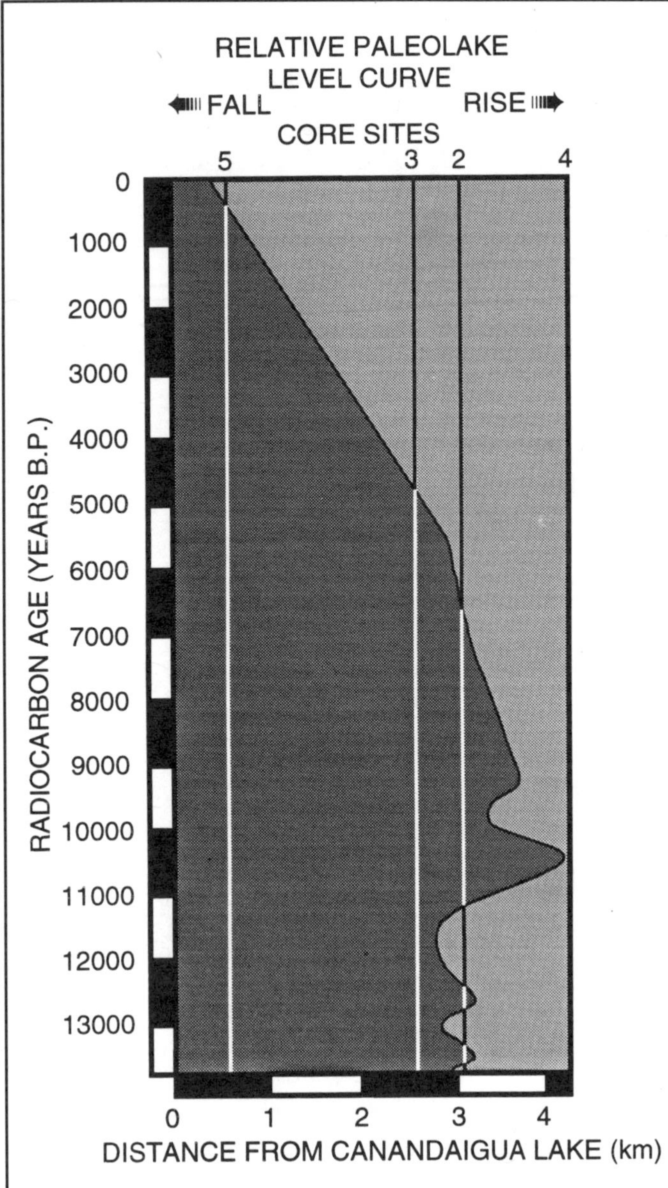

Figure 9. Relative lake-level curve over the last 13,650 B.P. for Canandaigua Lake based on facies architecture observed in cores (see Fig. 8).

and Shane, 1990), Ohio (Shane, 1987), New Jersey (Peteet et al., 1990), New England (Davis et al., 1980), as well as western New York (Spear and Miller, 1976). The dominance of oak in the middle to late Holocene suggests warmer temperatures (as well as reduced precipitation) or more evapotranspiration, or both (Shane, 1987). The reduced availability of moisture during the middle and late Holocene may explain the largely regressive lithostratigraphic changes observed in the Canandaigua Valley cores over the last ~8,000 B.P.

SUMMARY AND CONCLUSIONS

Sediment cores document the stratigraphic framework beneath the dry valley south of Canandaigua Lake (Fig. 1). The age control used in this study is based on both conventional and AMS radiocarbon dates (Table 1) supported by a magnetic record of secular variation of paleoinclination. The radiocarbon ages indicate variable sediment accumulation rates and confine the age of the late-glacial and postglacial record to less than 13,650 ± 210 B.P. (Fig. 4).

The lithostratigraphy beneath the dry lake valley south of Canandaigua Lake documents a series of rapid lake-level driven facies migrations (transgressions and regressions) (Figs. 8 and 9). Four millennium-scale (800 to 1,200 yr), lake-level fluctuations occurred between ~13,650 and 9,500 B.P., with relative highstand lake levels centered at ~13,500, 12,600, 10,400, and 9,400 B.P. (Fig. 9). Maximum relative lake level occurred at 10,400 B.P. when open lacustrine and marsh environments existed ~4.5 km south of modern-day Canandaigua Lake. After 9,400 B.P., Canandaigua Lake appears to have gradually fallen to its most regressed postglacial position.

The first-order vegetational changes at Canandaigua Lake are similar to those observed at other localities throughout northeastern North America and record the climatically driven succession of vegetation from a park–tundra–boreal forest (~13,400 to 9,900 B.P.), to a mixed hardwood–conifer forest (~9,900 to 9,645 B.P.), to a pine-dominated forest (~9,645 to 7,500 B.P.), and ultimately to a deciduous forest (<7,500 B.P.). Several of the significant vegetational changes observed at Canandaigua Lake include: (1) the abrupt and synchronous decline of deciduous and boreal taxa, and expansion of hemlock and alder between 11,300 and 10,200 B.P., which is similar to that observed elsewhere in northeastern North America, and adds to the growing documentation of colder and wetter conditions, likely in association with the Younger Dryas/Killarney events; (2) the establishment of a pine-dominated forest by 9,645 B.P., signaling the initiation of a climatic drying trend, in response to increased atmospheric heating at the beginning of the Holocene; (3) the establishment of an oak-dominated, deciduous forest by ~6,450 B.P., which correlates well to the Northern Hemisphere summer thermal maximum at ~6,000 B.P. (COHMAP Members, 1988); and (4) the decline of hemlock around Canandaigua Lake at ~4,800 B.P., which is similar to that observed elsewhere in northeastern North America (e.g., Allison et al., 1986), attesting to its usefulness as a time-stratigraphic horizon.

ACKNOWLEDGMENTS

This research was supported by National Science Foundation Grant (EAR–8703870) to H. T. Mullins. We thank Henry Mullins, Peter Plumley, Gene Domack, Don Cox, Bob Darling, Steve Good, William Anderson, and John Petruccione for their field and laboratory assistance; and Walter Dean and an anonymous reviewer for their comments.

REFERENCES CITED

Allison, T. D., Moeller, R. E., and Davis, M. B., 1986, Pollen in laminated sediments provides evidence for mid-Holocene forest pathogen outbreak: Ecology, v. 67, p. 1101–1105.

Ammann, B., and Lotter, A. F., 1989, Late glacial radiocarbon- and palynostratigraphy on the Swiss plateau: Boreas, v. 18, p. 109–126.

Anderson, T. W. and Lewis, C. F. M., 1985, Postglacial water-level history of the Lake Ontario basin, *in* Karrow, P. F., and Calkin, P. E., eds., Quaternary evolution of the Great Lakes: Geological Association of Canada Paper 30, p. 231–253.

Andrews, J. T., 1970, A geomorphological study of postglacial uplift with particular reference to arctic Canada: Institute of British Geographers Special Publication 2, 156 p.

Aravena, R., Warner, B. G., MacDonald, G. M., and Hanf, K. I., 1992, Carbon isotope composition of lake sediments in relation to lake productivity and radiocarbon dating: Quaternary Research, v. 37, p. 333–345.

Atkinson, T. C., Briffa, K. R., and Coope, G. R., 1987, Seasonal temperatures in Britian during the past 22,000 years, reconstructed using beetle remains: Nature, v. 325, p. 587–592.

Baker, R. G., Maher, L. J., Chumbley, C. A., and Van Zant, K. L., 1992, Patterns of Holocene environmental change in the midwestern United States: Quaternary Research, v. 37, p. 379–389.

Bernabo, J. C., and Webb, W. T., III, 1977, Changing patterns in the Holocene pollen record of northeastern North America: A mapped summary: Quaternary Research, v. 8, p. 64–96.

Brennan, S. F., Hamilton, M., Kilbury, R., Reeves, R. L., and Covert, L., 1984, Late Quaternary secular variation of geomagnetic declination in western New York: Earth and Planetary Science Letters, v. 70, p. 363–372.

Broecker, W. S., and Denton, G. H., 1989, The role of ocean-atmosphere reorganization in glacial cycles: Geochimica et Cosmochimica Acta, v. 53, p. 2465–2501.

Broecker, W. S., Andree, M., Wolfi, W., Oescheger, H., Bonani, G., Kennett, J., and Peteet, D., 1988, The chronology of the last deglaciation: Implications for the cause of the Younger Dryas event: Paleooceanography, v. 3, p. 1–19.

Broecker, W. S., Bond, G., and Klas, M., 1990, A salt oscillator in the glacial Atlantic? I. The concept: Paleoceanography, v. 5, p. 469–477.

Calkin, P. E., and Muller, E. H., 1992, Pleistocene stratigraphy of the Erie and Ontario Lake Bluffs in New York, *in* Quaternary coasts of the United States: Marine and Lacustrine Systems, Society of Economic Petrologists and Mineralogists Special Publication 48, p. 385–396.

Carmichael, C. M., Mothersill, J. S., and Morris, W. A., 1990, Paleomagnetic and pollen chronostratigraphic correlations of the late-glacial and postglacial sediments in Lake Ontario: Canadian Journal of Earth Sciences, v. 27, p. 131–147.

Clark, R. H., and Personage, N. P., 1970, Some implications of crustal movement in engineering planning: Canadian Journal of Earth Sciences, v. 7, p. 628–633.

Coates, D. R., 1974, Reappraisal of the glaciated Appalachian Plateau, *in* Coates, D. R., ed., Glacial geomorphology, Proceedings Volume, Fifth Annual Geomorphology Symposia Series: Binghamton, New York, p. 205–243.

COHMAP Members, 1988, Climatic changes of the last 18,000 years: Observations and model simulations: Science, v. 241, p. 1043–1052.

Cox, D. D., 1959, Some post-glacial forests in central New York: Ecology, v. 20, p. 264–271.

Cox, D. D. and Lewis, D. M., 1965, Pollen studies in the Crusoe Lake area of prehistoric Indian occupation: New York State Museum Science Service, v. 387, 29 p.

Creer, K. M., and Tucholka, P., 1982, Construction of type curves of geomagnetic secular variation for dating lake sediments from east central North America: Canadian Journal of Earth Sciences, v. 19, p. 1106–1115.

Dalrymple, R. W., and Carey, J. S., 1990, Water level fluctuations in Lake Ontario over the last 4,000 years as recorded in the Catarqui River lagoon, Kingston, Ontario: Canadian Journal of Earth Sciences, v. 27, p. 1330–1338.

Dansgaard, W, Clausen, H. S., Guendestrup, N., Hammer, C. U., Johnsen, S. J., Kristinsdottir, P. M., and Reeh, N., 1982, A new deep Greenland ice core: Science, v. 218, p. 1273–1277.

Dansgaard, W., White, J. W. C., and Johsen, S. J., 1989, The aburpt termination of the Younger Dryas event: Nature, v. 339, p. 532–533.

Davis, M. B., 1969, Climatic changes in southern Connecticut recorded by pollen deposition at Roger Lake: Ecology, v. 50, p. 409–422.

Davis, M. B., 1981, Outbreaks of forest pathogens in forest history, *in* Proceedings, Fourth International Palynological conference, Birbal Sahni Institute of Paleobotany: Lucknow, India, v. 3, p. 216–277.

Davis, M. B., 1983, Holocene vegetational history of the eastern United States, *in* Wright, H. E., Jr., ed., Late Quaternary environments of the United States: Minneapolis, University of Minnesota Press, p. 166–181.

Davis, M. B., Spear, R. W., and Shane, L. C. K., 1980, Holocene climate of New England: Quaternary Research, v. 14, p. 240–250.

Dean, W. E., Jr., 1974, Determination of carbonate and organic matter in calcareous sediments and sedimentary rocks by loss-on-ignition: Comparison with other methods: Journal of Sedimentary Petrology, v. 44, p. 242–248.

Deevy, E. S., Jr., 1943, Studies on Connecticut lake sediments. I. A postglacial climatic chronology from southern New England: American Journal of Science, v. 237, p. 691–723.

Denton, G. H., and Karlen, W., 1973, Holocene climatic variability and their possible cause: Quaternary Research, v. 3, p. 155–205.

Duplessy, J. C., Delibrias, G., Turon, J. L., Pujol, C., and Dupart, J., 1981, Deglacial warming of the northeastern North Atlantic Ocean: Correlation with the palaeoclimatic evolution of the European continent: Palaeogeography, Palaeoclimatology, Palaeoecology, v. 35, p. 121–144.

Dwyer, T. R., and Darling, R., 1993, Deglacial lake level fluctuations at Canandaigua Lake, NY: Northeastern Section, Geological Society of America Abstracts with Programs, v. 25, p. 13.

Eaton, S. W., and Kardos, L. P., 1978, The limnology of Canandaigua Lake, *in* Bloomfield, J. A., ed., Lakes of New York state. Vol. 1, Ecology of the Finger Lakes: New York, Academic Press, p. 226–307.

Eicher, U., 1980, Pollen-und Sauerstoffistopenanalysen an spatglazialen profilen vom Grezensee, Faulenseemoos und vom Regenmoosob: Boltigen Mitt. Naturforschungs Gesellschaft Bern, v. 37, p. 65–80.

Faegri, K., and Iversen, I., 1975, Textbook of pollen analysis: New York, Hafner Press, 295 p.

Flint, J. E., Dalrymple, R. W., and Flint, J. J., 1988, Stratigraphy of the Sixteen Mile Creek lagoon, and its implications for Lake Ontario water levels: Canadian Journal of Earth Sciences, v. 25, p. 1175–1183.

Flohn, H., 1986, Singular events and catastrophes now and in climatic history: Naturwissenschaften, v. 73, p. 136–149.

Fritz, P., Morgan, A. V., Eicher, U., and McAndrews, J. H., 1987, Stable isotope, fossil coleoptera, and pollen stratigraphy in Late Quaternary sediments from Ontario and New York state: Palaeogeography, Palaeoclimatology, Palaeoecology, v. 58, p. 183–202.

Fullerton, D. S., 1986, Stratigraphy and correlation of glacial deposits from Indiana to New York and New Jersey, *in* Sibrava, V., Bowen, D. Q., and Richmond, G. M., eds., Quaternary glaciations in the Northern Hemisphere: New York, Pergamon Press, p. 23–37.

Good, S. C., 1987, Mollusc-based interpretations of lacustrine paleoenvironment of the Sheep Pass Formation (Latest Cretaceous to Miocene) of east central Nevada: Palaios, v. 2, p. 467–478.

Hansel, A. K., and Mickelson, D. M., 1988, A reevaluation of the timing and cause of high lake phases in the Lake Michigan basin: Quaternary Research, v. 29, p. 113–128.

Keen, K. L., and Shane, L. C. K., 1990, A continuous record of Holocene eolian activity and vegetation change at Lake Ann, east-central Minnesota: Geological Society of America Bulletin, v. 102, p. 1646–1657.

King, J. W., Banerjee, S. K., Marvin, J., and Lund, S., 1983, Use of small amplitude paleomagnetic fluctuations for correlation and dating of continental climatic changes: Palaeogeography, Palaeoclimatology, Palaeoecology, v. 42, p. 167–183.

Krall, D. B., 1977, Late Wisconsinan ice recession in east-central New York: Geological Society of America Bulletin, v. 88, p. 1697–1710.

Leopold, E. B., 1956, Two late-glacial deposits in southern Connecticut: Proceedings, National Academy of Sciences (U.S.), v. 52, p. 863–867.

Levesque, A. J., Mayle, F. E., Walker, I. R., and Cwynar, L. C., 1993. A previously unrecognized late-glacial cold event in eastern North America: Nature, v. 361, p. 623–626.

Lister, G. S., 1987, A 15,000 year isotopic record from Lake Zurich of deglaciation and climatic change in Switzerland: Quaternary Research, v. 29, p. 129–141.

Lowe, J. J., Lowe, S., Fowler, A., Hedges, R. E. M., and Austin, J., 1988, Comparison of accelerator and radiometric measurements obtained from late Devensian late glacial lake sediments from Llyn Gwernan, North Wales, U.K.: Boreas, v. 17, p. 335–369.

Mathewes, R. F., Hensen, L. E., and Patterson, R. T., 1993, Evidence for a younger Dryas-like cooling event on the British Columbia coast: Geology, v. 21, p. 101–104.

McCulloch, W. F., 1939, A post glacial forest in central New York: Ecology, v. 20, p. 264–271.

Miller, N. G., 1973, Late glacial plants and plant commumities in northwestern New York state: Journal of the Arnold Arboretum, v. 54, p. 123–159.

Miller, N. G., 1988, The late Quaternary Hiscock site, Genesee County, New York: Paleoecological studies based on pollen and plant macrofossils, *in* Laub, R. S., Miller, N. G., and Steadman, D. W., eds., Late Pleistocene and Early Holocene paleoecology and archeology of the eastern Great Lakes region: Bulletin of the Buffalo Society of Natural Sciences, v. 33, p. 83–93.

Mott, R. J., Grant, D. R., Stea, R., and Ochietti, S., 1986, Late-glacial climatic oscillations in Atlantic Canada equivalent to Allerod/Younger Dryas event: Nature, v. 123, p. 247–250.

Muller, E. H., and Calkin, P. E., 1993, Timing of Pleistocene glacial events in New York state: Canandian Journal of Earth Sciences, v. 30, p. 1829–1845.

Muller, E. H., and Prest, V. K., 1985, Glacial lakes in the Ontario Basin, *in* Karrow, P. F., and Calkin, P. E., eds., Quaternary evolution of the Great Lakes: Geological Association of Canada Paper 30, p. 211–229.

Mullins, H. T., and Hinchey, E. J., 1989, Erosion and infill of New York Finger Lakes: Implications for Laurentide Ice Sheet deglaciation: Geology, v. 17, p. 622–625.

Oescheger, H., Beer, J., Seigenthaler, U., Stauffer, B., Dansgaard, W., and Langway, C. C., 1984, Late glacial climate history from ice cores, *in* Hensen, J., and Takahashi, T., eds., Climate processes and climate sensitivity: American Geophysical Union Monograph 29, p. 299–306.

Ogden, J. G., III, 1967, Evidence for a sudden climatic change around 10,000 years ago, *in* Cushing, E. J., and Wright, H. E., Jr., eds., Quaternary Paleoecology: New Haven, Connecticut, Yale University Press, 425 p.

O'Rourke, M. K., 1976, An absolute pollen chronology of Seneca Lake, N.Y. [M.S. thesis]: Tucson, University of Arizona, 82 p.

Overpeck, J. T., 1985, A pollen study of a late Quaternary peat bog south-central Adrirondack Mountains, New York: Geological Society of America Bulletin, v. 96, p. 145–154.

Overpeck, J. T., 1991, Century to millennium-scale climatic variability during the late Quaternary, *in* Bradley, R. S., ed., Global change of the past: Boulder, Colorado, University Corporation for Atmospheric Research, Office for Interdisciplinary Earth Studies, p. 139–173.

Peltier, W. R., 1987, Glacial isostasy, mantle viscosity, and Pleistocene climate change, *in* Ruddiman, W. F., and Wright, H. E., eds., North America and adjacent oceans during the last deglaciation: Boulder, Colorado, Geological Society of America, Geology of North America, v. K–3, p. 155–182.

Peltier, W. R., and Andrews, J. T., 1976, Glacial isostatic adjustment. I. The forward problem: Geophysical Journal of the Royal Astronomical Society, v. 46, p. 605–646.

Peteet, D. M., Vogel, J. S., Nelson, D. E., Southon, J. R., Nickmann, R. J., and Heusser, L. E., 1990, Younger Dryas climatic reversal in northeastern USA? AMS ages for an old problem: Quaternary Research, v. 33, p. 219–230.

Peteet, D. M., Daniels, R. A., Heusser, L. E., Vogel, J. S., Southan, J. R., and Nelson, D. E., 1993, Late-glacial pollen, macrofossils, and fish remains in northeastern USA—The Younger Dryas oscillation: Quaternary Science Reviews, v. 12, p. 597–612.

Prest, V. K., 1970, Quaternary geology of Canada, *in* Geology and Economic Minerals of Canada: Geologic Survey of Canada: Economic Geology Report 1, 5th Ed., p. 675–764.

Rawlence, D. J., and Senior, A., 1988, A late-glacial diatom and pigment history of Little Lake, New Brunswick, with particular reference to the Younger Dryas climatic oscillation: Journal of Paleolimnology, v. 1, p. 163–177.

Ridge, J. C., Brennan, S. F., and Muller, E. H., 1990, The use of declination to test correlations of late Wisconsinan glaciolacustrine sediments in central New York: Geological Society of America Bulletin, v. 102, p. 26–44.

Ruddiman, W. F., 1987, Synthesis: The oceanic/ice sheet record, *in* Ruddiman, W. F., and Wright, H. E., Jr., eds., North America and adjacent oceans during the last deglaciation: Boulder, Colorado, Geological Society of America, Geology of North America, v. K–3, p. 463–478.

Ruddiman, W. F., and McIntyre, A., 1981, Oceanic mechanisms for the amplification of the 23,000 year ice-volume cycle: Science, v. 212, p. 617–627.

Shane, L. C. K., 1987, Late-glacial vegetational and climatic history of the Allegheny Plateau and the Till Plains of Ohio and Indiana, USA: Boreas, v. 16, p. 1–20.

Spear, R. W., and Miller, N. G., 1976, A radiocarbon dated pollen diagram from the Allegheny Plateau of New York State: Journal of the Arnold Arboretum, v. 57, p. 369–403.

Street-Perrott, F. A., and Harrison, S. P., 1984, Temporal variations in lake levels since 30,000 yr. BP—An index of global hydrological cycle, *in* Hensen, J., and Takahashi, T., eds., Climate processes and climate sensitivity: American Geophysical Union Monograph 29, p. 118–129.

Tarr, R. S., 1904, Glacial erosion in the Finger Lakes region of central New York: Journal of Geology, v. 14, p. 18–21.

Tinkler, K. J., Pengelly, J. W., Parkins, W. G., and Terasmae, J., 1992, Evidence of high water level in the Erie basin during the Younger Dryas chronozone: Journal of Paleolimnology, v. 7, p. 215–234.

Walcott, R. I., 1972, Late Quaternary vertical movements in eastern North America: Reviews in Geophysics and Space Physics, v. 10, p. 849–884.

Walker, I. R., Mott, R. J., and Smol, J. P., 1991, Allerod–Younger Dryas lake temperatures from midge fossils in Atlantic Canada: Science, v. 253, p. 1010–1012.

Watts, W. A., 1980, Regional variations in the response of vegetation to late-glacial climatic events in Europe, *in* Lowe, J. J., Gray, J. M., and Robinson, J. E., eds., The late glacial of north-west Europe: New York, Pergamon Press, 205 p.

Webb, T., III, Bartlein, P. J., and Kutzbach, J. E., 1987, Climatic change in eastern North America during the past 18,000 years: Comparisons of pollen data with model results, *in* Ruddiman, W. F., and Wright, H. E., Jr., eds., North America and adjacent oceans during the last deglaciation: Boulder, Colorado, Geological Society of America, Geology of North America, v. K–3, p. 447–462.

Welten, M., 1982, Vegetationsgeschichtliche Unterschungen in den Westlichen Schweizer Alpen: Bern-Wallis: Denkschrift Schweizerische Naturforschungsgesellschaft, v. 95, p. 1–104.

Weninger, J. M., and McAndrews, J. H., 1989, Late Holocene aggradation in the lower Humber River valley, Ontario: Canadian Journal of Earth Sciences, v. 26, p. 1842–1844.

Whitehead, D. R., Charles, D. F., Jackson, S. T., Smol, J. P., and Engstrom, D. R., 1989, The developmental history of Adirondack (N.Y.) Lakes: Journal of Paleolimnology, v. 2, p. 185–206.

Wright, H. E., Jr., 1968, History of the prairie peninsula, *in* Bergstrom, R. E., ed., The Quaternary of Illinois, University of Illinois College of Agriculture Special Publication 14, p. 78–88.

Wright, H. E., Jr., 1989, The amphi-Atlantic distribution of the younger Dryas paleoclimatic oscillation: Quaternary Science Reviews, v. 8, p. 295–306.

Manuscript Accepted by the Society January 16, 1996

Printed in U.S.A.

Geological Society of America
Special Paper 311
1996

Seismic reflection investigation of Montezuma wetlands, central New York State: Evolution of a Late Quaternary subglacial meltwater channel system

John L. Petruccione* and Robert W. Wellner*
Department of Earth Sciences, Heroy Geology Laboratory, Syracuse University, Syracuse, New York 13244
Robert E. Sheridan
Department of Geological Sciences, Wright-Reiman Geological Laboratory, Rutgers University, New Brunswick, New Jersey 08903

ABSTRACT

Montezuma wetlands occupy a large, relict, southward-directed drainage system that funnels into the north end of Cayuga Lake in the central Finger Lakes region of New York State. The origin of this distinct channel system has not been addressed despite the fact that it is a prominent feature of the drumlinized Ontario Lowland, where today the drainage is directed northward.

This study investigates the near-surface geology of the Montezuma channel system in order to evaluate its origin and regional significance. Approximately 12 km of land-based multichannel seismic reflection data were acquired from Montezuma channels, analyzed, and correlated with water-well, gas-well, and test boring data. Results document a broad, shallow, channelized erosion surface beneath the wetlands that slopes southward in an undulatory fashion. This erosion surface extends at least 56 m beneath local base level before rising over a bedrock sill near the north end of Cayuga Lake. Seismic reflection profiles from the northern end of Cayuga Lake reveal a northward-thickening sediment wedge (up to 60 m) that extends south of the Montezuma channels.

These data suggest that the Montezuma channels were eroded into late Quaternary sediments and underlying Paleozoic bedrock by southward-flowing, subglacial meltwaters issuing from beneath the southern margin of the Laurentide ice sheet. Subglacial meltwater appears to have been focused and ultimately channelized into this region of relatively low surface elevation sometime between 14.4 and 13.6 ka. Following ice sheet withdrawal and deposition of coarse-grained channel fill, pro- and postglacial lakes occupied the low-lying Montezuma channels, further infilling them with fine-grained, lacustrine sediments. Underfit, northward-directed drainage through the wetlands today bears little resemblance to the south-flowing, subglacial meltwaters that appear to have formed the Montezuma channels during deglaciation.

*Present addresses: Petruccione, Geophysics Division, Berkshire Environmental Associates, Inc., 409 Penn Avenue, Sinking Spring, Pennsylvania 19608-1109; Wellner, Exxon Production Research Company, P.O. Box 2189, Houston, Texas 77252-2189.

Petruccione, J. L., Wellner, R. W., and Sheridan, R. E., 1996, Seismic reflection investigation of Montezuma wetlands, central New York State: Evolution of a Late Quaternary subglacial meltwater channel system, *in* Mullins, H. T., and Eyles, N., eds., Subsurface Geologic Investigations of New York Finger Lakes: Implications for Late Quaternary Deglaciation and Environmental Change: Boulder, Colorado, Geological Society of America Special Paper 311.

INTRODUCTION

The Montezuma wetlands consist of a southward-directed, channelized drainage system (Fig. 1) that funnels into the north end of Cayuga Lake, one of the two largest Finger Lakes of central New York State. This area occupies approximately 125 km^2 and is one of the largest freshwater wetlands in New York State. The Montezuma channels extend approximately 25 km north of Cayuga Lake with a maximum east-west width of 6 km. In addition to the Finger Lakes, the morphology of the wetlands is one of the most distinctive features of the region, yet one of the least studied (von Engeln, 1961; Coates, 1968).

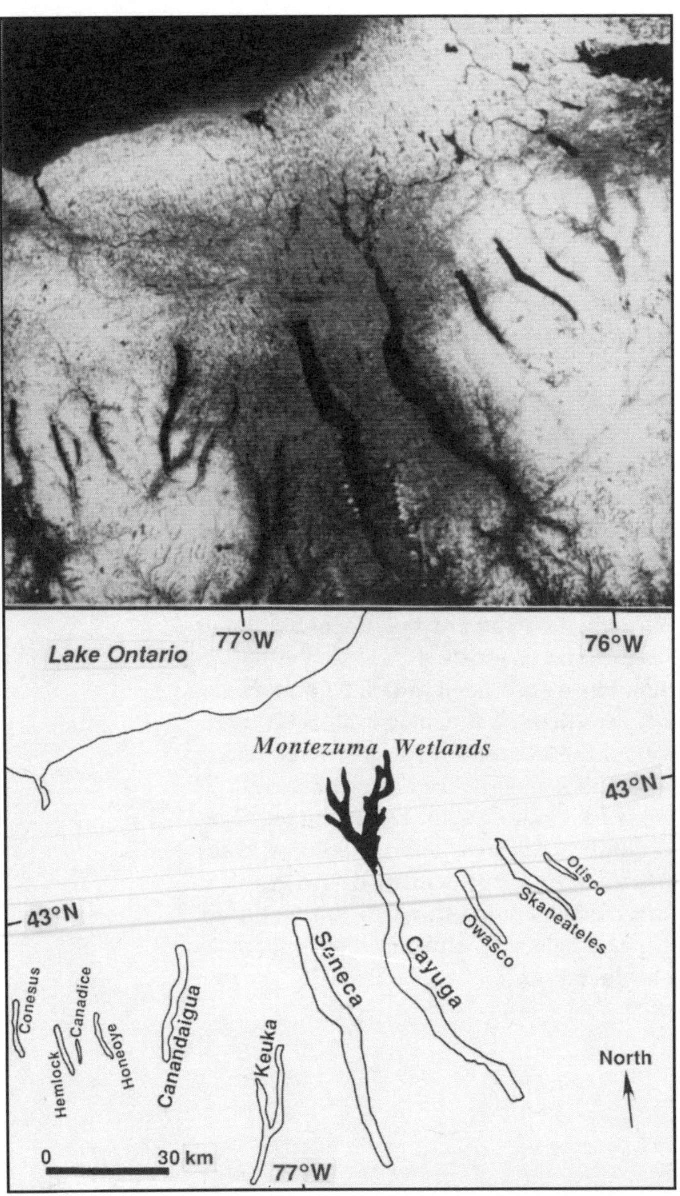

Figure 1. Satellite image (U.S. Geological Survey, EROS Data Center, Sioux Falls, South Dakota, no. E-1234–15244–502) of the Finger Lakes region (top) and map of the 11 Finger Lakes highlighting the southward-directed drainage pattern of the Montezuma wetlands (bottom).

The Finger Lakes (Fig. 1) occur along the northern margin of the glaciated Appalachian Plateau (Coates, 1974) delineated by the Onondaga Escarpment, an extensive, east-west–trending outcrop of resistant Devonian limestone. This plateau descends northward via a series of steps to the relatively flat, but drumlinized, Ontario Lowlands that extend north of the Finger Lakes, to Lake Ontario.

Fairchild (1934, p. 235) described Cayuga Lake Valley as "the axial depression of central New York," as it is the topographically lowest area (lake level datum, 118 m) in the Finger Lakes region (Bloom, 1986). A high-resolution seismic reflection investigation of Cayuga Lake (Mullins and Hinchey, 1989; Mullins et al., 1991) has revealed an axially overdeepened bedrock trough beneath the lake eroded as much as 242 m *below* sea level. Based on seismic stratigraphic analysis, the late Quaternary sediments within Cayuga Lake appear to be the result of a single late Wisconsin infill event, deposited between ~14.4 and 13.6 ka* (Mullins and Hinchey, 1989; Mullins et al., 1991).

SURFICIAL GEOLOGY

The Montezuma channels occur along the southern margin of a well-developed drumlin field (Fairchild, 1907), that characterizes the Ontario Lowlands. North-south flowline patterns exhibited by the drumlins are believed to be the result of either outlet glacier flow from an ice dome centered over Lake Ontario (Ridky and Bindschadler, 1990) or the result of a phase of rapid flow of the Laurentide ice sheet (White, 1985). Immediately south of the Montezuma channels are a series of till moraines (Muller and Cadwell, 1986), which likely represent ice margin positions of the retreating Laurentide ice sheet (Fig. 2).

The majority of the modern-day Montezuma wetlands is swamp underlain by deposits of peat, marl, and organic-rich silt/sand in poorly drained localities (Fig. 2) (Muller and Cadwell, 1986). The channel system bifurcates into eastern and western branches approximately 6 km north of Cayuga Lake (Figs. 2 and 3). The western channel narrows in width as it extends northward from Cayuga Lake, whereas the eastern branch of the system broadens to impound a series of higher elevation drumlin "islands" before gradually narrowing to the north (Fig. 3). Present-day drainage by underfit, meandering streams is northward with much of the free-standing water confined to shallow, low-lying pools and wetland areas. The primary purpose of this study is to determine the extent of erosion and thickness of unconsolidated sediment-fill beneath the Montezuma wetlands, and to evaluate these data in terms of a model for the origin and evolution of the Montezuma wetlands.

METHODS

Approximately 12 km of land-based multichannel seismic reflection data were collected along roadways and artificial levees during the summers of 1988 and 1990. Data were acquired

*All ages referred to in this chapter are in uncalibrated radiocarbon years before present (1950).

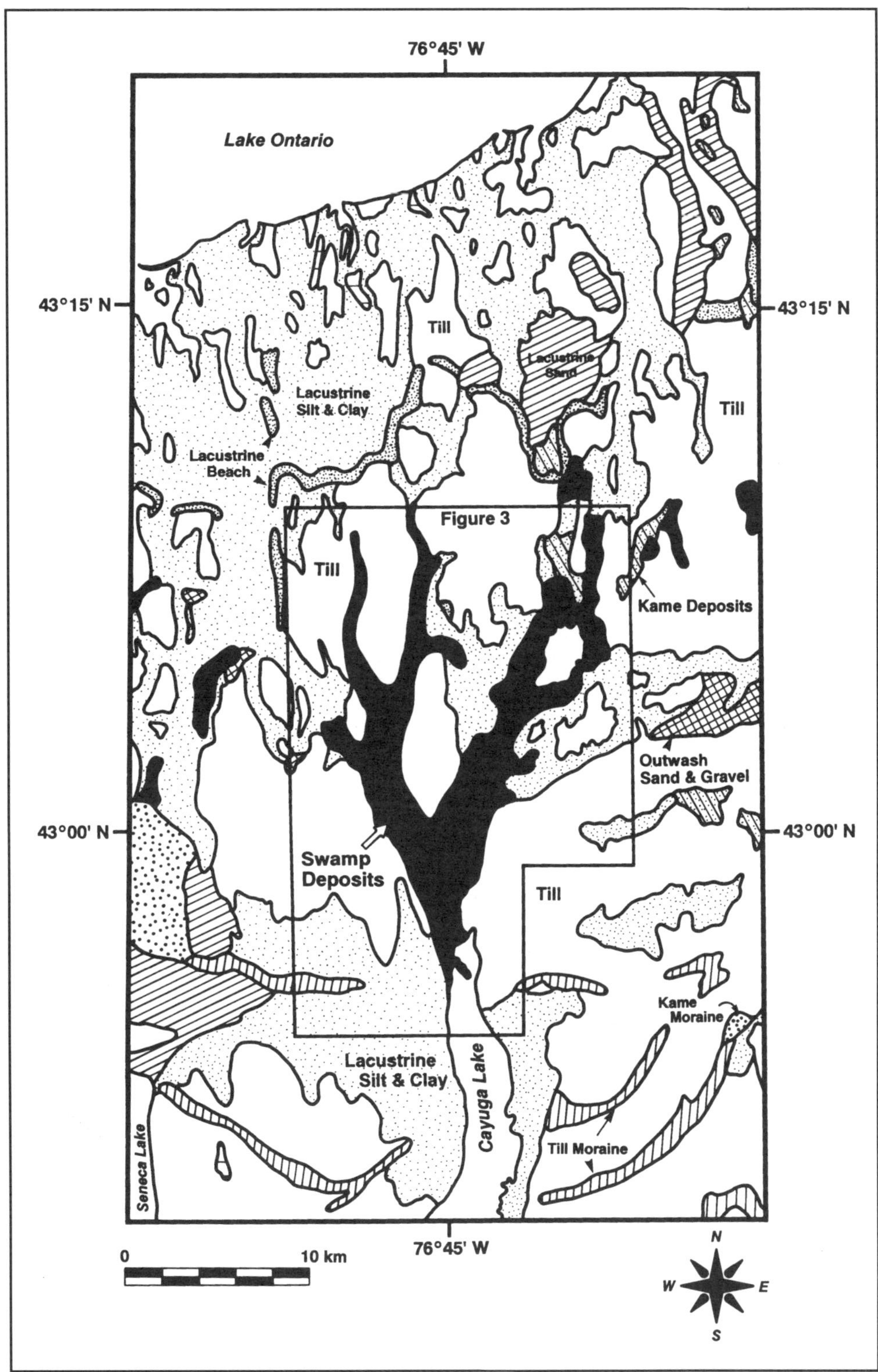

Figure 2. Surficial geologic map of Montezuma wetlands and northern Cayuga Lake region (adapted from Muller and Cadwell; 1986). Modern wetlands are designated as swamp deposits and highlighted in black.

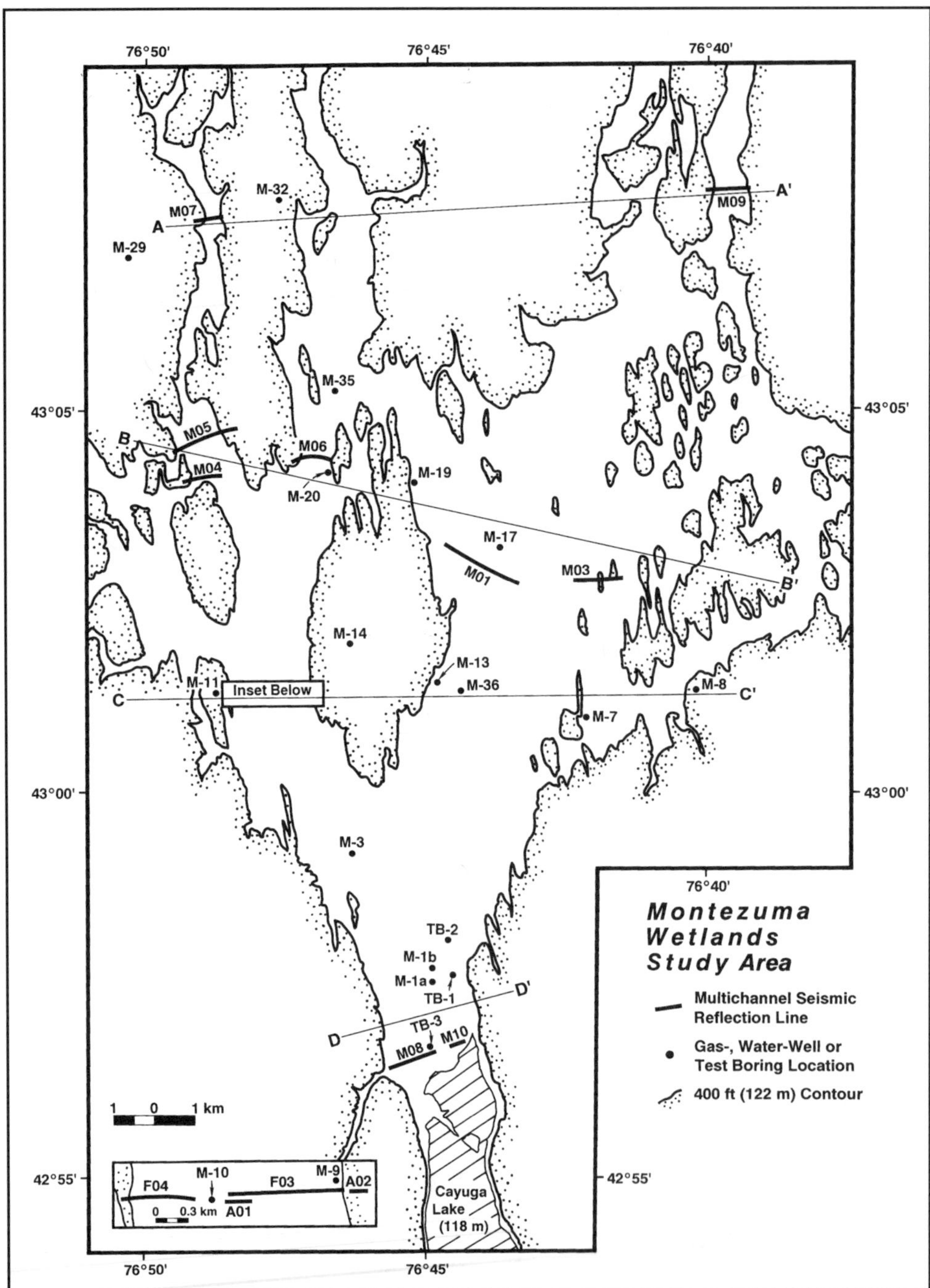

Figure 3. Detailed location map of Montezuma study area illustrating location of seismic reflection lines, water wells and gas wells, test borings, and cross-sectional profiles (A-A′ to D-D′) illustrated in Figure 13. Wetland areas are defined by 400-ft (122 m) contour, with stippled pattern lying above 400-ft contour.

using a 12-channel BISON GeoPro 8012A seismograph. A 230-kg weight-bag dropped from a height of 3 m served as the sound source. Shotpoint spacing was 20 m, with a total seismic spread coverage of 240 m.

Additional seismic reflection data were collected during May 1991 (0.4 km) using a sledgehammer (4.5 kg) as a sound source (e.g., Miller et al., 1986). Geophone spacing was 3 m and designed to yield higher resolution imaging of the sediment/bedrock interface (Hunter et al., 1984; Miller et al., 1990). A variable offset (15 to 24 m) was determined in the field by a walkaway noise test (e.g., Burger, 1989) in order to best locate the bedrock reflection window (Fig. 4). Processing of the multifold field records is described in Petruccione (1994).

In order to determine sediment thickness, acoustic travel

time is normally converted to depth using compressional wave velocities. A velocity scanning technique provided an objective method to choose stacking velocities at known time designations (Hodgson, 1984; Dobrin and Savit, 1988; Somanas et al., 1988; Li, 1991). Interval velocities (Dix, 1955) were determined using a velocity scanning program and ranged between 1,000 m/sec (unsaturated sediments) and 1,750 m/sec (saturated sediments). An average compressional wave velocity of 1,550 m/sec for the unconsolidated sediments within the Montezuma wetlands was selected for all time to depth conversions. The water wells and gas wells (unpublished U.S. Geological Survey database, Ithaca, New York; Cox and Lewis, 1965; Kantrowitz, 1970), as well as test borings (W. M. Kappel, U.S. Geological Survey, Ithaca, New York, personal communication, 1994) used to correlate bedrock to seismic sections and evaluate the stratigraphic nature of the sediment fill, are located in Figure 3 and illustrated in Figure 5.

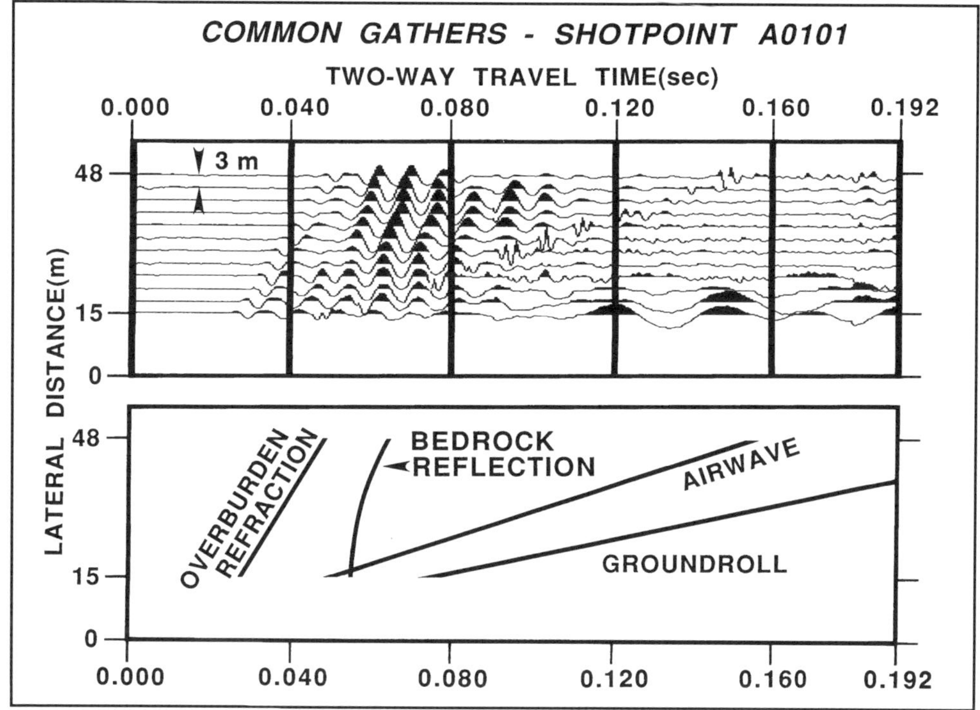

Figure 4. Twelve-channel (sledgehammer source) shotpoint record illustrating the bedrock reflection window.

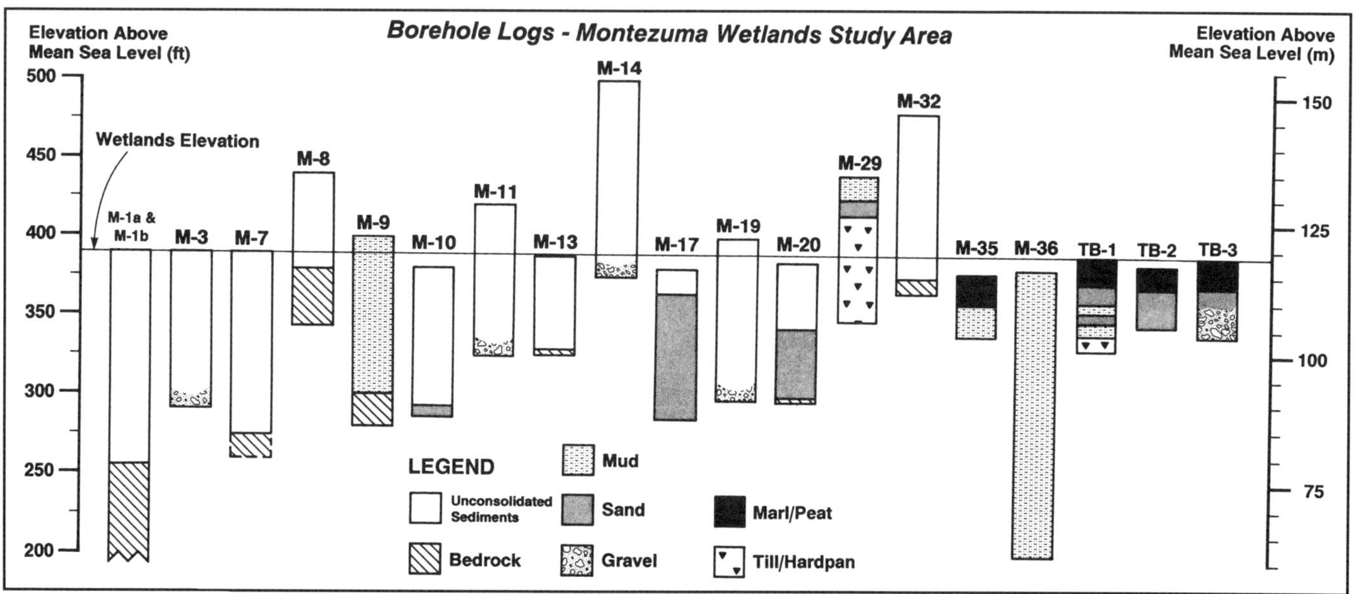

Figure 5. Water-well, gas-well, and test boring data that document depth to bedrock and infill thickness within the Montezuma wetlands study area (unpublished U.S. Geological Survey database, Ithaca, New York; Cox and Lewis, 1965; Kantrowitz, 1970).

RESULTS

Southern Montezuma wetlands

Two seismic reflection lines (M08 and M10; Fig. 3), covering a lateral distance of 1.5 km, were collected on a wetlands access road approximately 1 km north of Cayuga Lake. Exploratory (total depth, 215 m) gas wells (M-1a and M-1b; Fig. 3), drilled in the Montezuma National Wildlife Refuge approximately 2 km north of lines M08 and M10, intersected the top of bedrock at a subsurface depth of 42 m (Fig. 5). Based on a high-amplitude reflector at 0.05-sec two-way travel time (TWT), and using a compressional wave velocity of 1,550 m/sec, the top of bedrock extends approximately 40 m beneath seismic line M10. U.S. Geological Survey test boring (TB-3, Fig. 3) is located on seismic line M08 and was drilled to a subsurface depth of 18 m. This boring did not penetrate bedrock, terminated in sandy gravel (~12 m thick), and is overlain by approximately 6 m of peat and marl (Fig. 5).

Bedrock beneath seismic line M08 occurs at a maximum of 46 m below the land surface (73 m above mean sea level). This depth is considerably shallower than the depth to bedrock documented by Mullins and Hinchey (1989), who reported that bedrock extends as much as 75 m beneath the northern portion of Cayuga Lake basin (~10 km south of line M08, Fig. 2). Well M-3 (Fig. 3), located 5.5 km north of line M08, did not reach bedrock and terminated in sand and gravel at a depth of 31 m (Fig. 5).

Kantrowitz (1970) reported that 56 m of unconsolidated, fine-grained sediments were drilled at well M-36 (Fig. 5) and that bedrock was *not* encountered. Well M-36 is located approximately 7 km north of gas wells M-1a and M-1b (Fig. 3) and is at the same elevation as Cayuga Lake (118 m). The erosion surface thus appears to deepen northward by at least 10 m, based on these well and seismic reflection records, and a bedrock high (or sill) exists at the south end of the Montezuma wetlands.

Central Montezuma wetlands

Seismic lines F03 and F04. Seismic reflection lines F03 and F04 covered 2 km of the western branch of Montezuma wetlands, approximately 10 km north of Cayuga Lake (inset, Fig. 3). A strong, high-amplitude reflector at approximately 0.04- to 0.05-sec TWT defines the sediment/bedrock contact (Fig. 6). Laterally continuous reflections downlap bedrock along the central and western portions of the profile, whereas reflection-free facies occur across the uppermost portions (0.03-sec TWT) of seismic lines F03 and F04.

Maximum depth to bedrock is 38 m and is located within a bedrock channel along the west-central portion of seismic line F04 (Fig. 6). U.S. Geological Survey test well M-10 bottomed in sand at a subsurface depth of 27 m (Fig. 5); these sediments correlate with the higher amplitude reflections on lines F03 and F04. Finer grained sediments (clay and silt) appear to overlie coarser grained channel-fill sediments (sand and gravel) based on drill records as well as on seismic facies relationships. Well M-9 (Fig. 5), however, recorded 31 m of only clay and silt that rest directly on bedrock (unpublished, U.S. Geological Survey database, Ithaca, New York).

Seismic line A01. Seismic reflection line A01 (~0.2 km) was collected parallel to the western end of line F03 (inset, Fig. 3), using a sledgehammer source and tighter geophone spacing, which resulted in an enhanced bedrock reflection. A laterally continuous, high-amplitude bedrock reflection was recorded at approximately 0.04-sec TWT (Fig. 7). Strong, high-amplitude reverberations underlie and mimic this primary reflection due to the high acoustic impedance contrast between unconsolidated sediments and bedrock. Bedrock occurs at a maximum depth of 35 m along the western edge of seismic line A01, which is in good agreement with the weightdrop seismic data of line F03 (Fig. 6). The bedrock surface dips slightly to the west (Fig. 7), which is consistent with the bedrock channel morphology defined by seismic lines F03 and F04 (Fig. 6).

Seismic line A02. Approximately 0.1 km east of line F03,

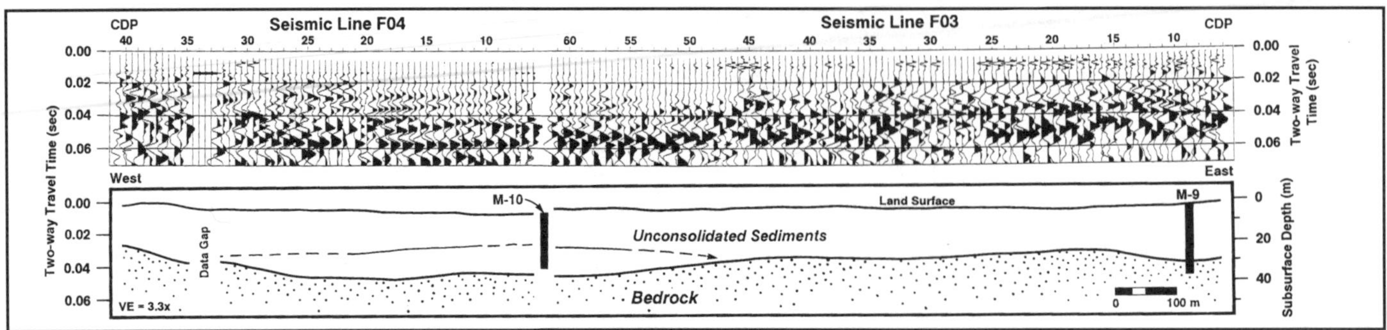

Figure 6. Weight-drop, multichannel seismic reflection profile F04 and F03 (top) and line drawing interpretation (bottom) collected across the western branch of the Montezuma wetlands. Note high-frequency reflection character of unconsolidated sediments that overlie channelized surface along west-central portion of seismic profile. Depth scale for all seismic profiles calculated using a P-wave velocity of 1,550 m/sec. CDP = common depth point.

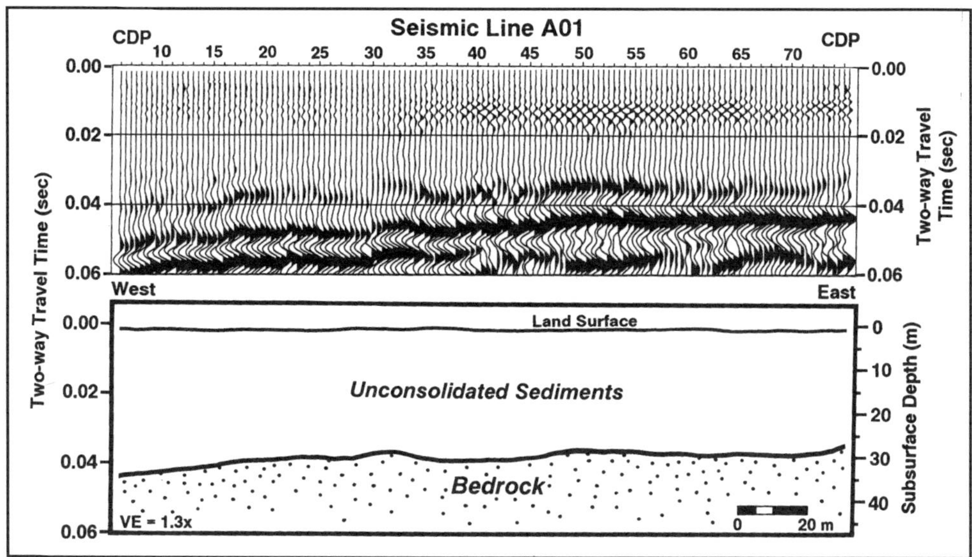

Figure 7. Sledgehammer, multichannel seismic reflection profile A01 (top), and line drawing interpretation (bottom) acquired across the central portion of the Montezuma wetlands' western branch.

line A02 (0.15 km) was acquired over the crest of a drumlin (inset, Fig. 3) using a sledgehammer sound source and 3 m geophone spacing. A strong, laterally continuous, high-amplitude bedrock reflection is well-defined between 0.03- and 0.04-sec TWT (Fig. 8). Depths to bedrock range from 23 m at the drumlin crest to a maximum depth of 30 m along flanking troughs. Calculated relief on the bedrock surface (trough to crest) is 12 m, and surface topography appears to be controlled by bedrock.

Seismic lines M01 and M03. Seismic reflection profiles M01 and M03 cover a lateral distance of 3 km and are located approximately 11 km north of Cayuga Lake (Fig. 3). These reflection profiles are characterized by an upper transparent facies with an average thickness of 25 m (Fig. 9). Below the

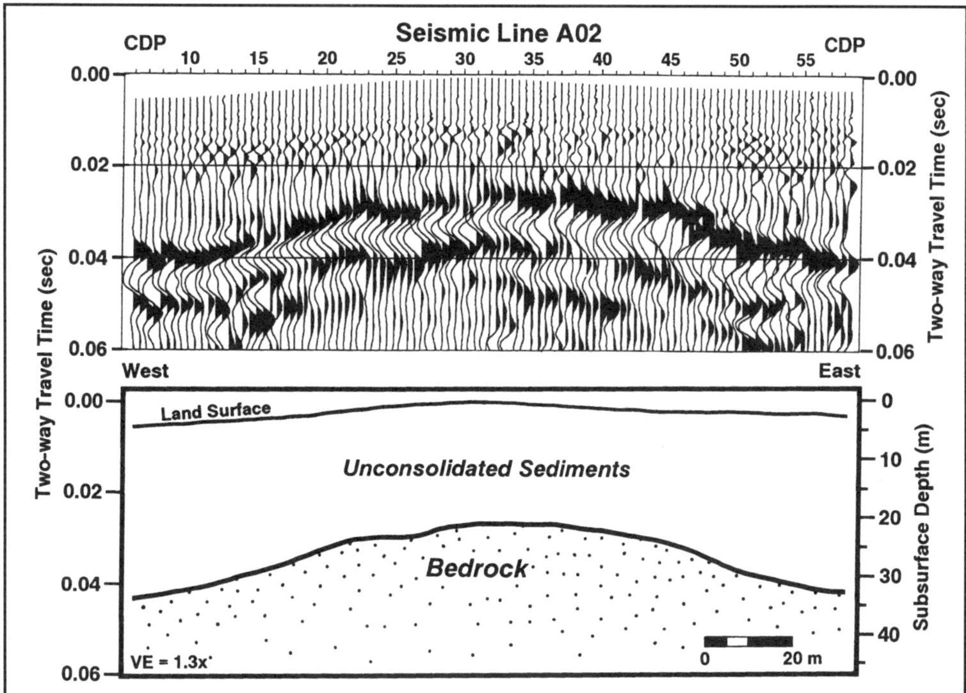

Figure 8. Sledgehammer, multichannel seismic reflection profile A02 (top), and line drawing interpretation (bottom) collected across a drumlin bordering the western channel of Montezuma wetlands. Surface topography appears to mimic a bedrock high beneath drumlin.

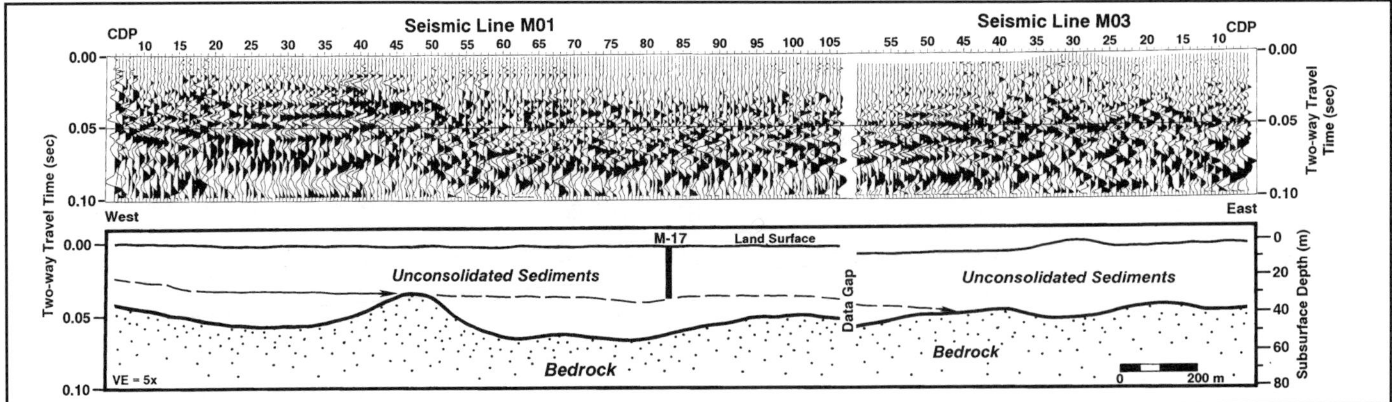

Figure 9. Weight-drop, multichannel seismic reflection profile M01 and M03 (top), and line drawing interpretation (bottom) collected across the eastern branch of the Montezuma channels. High-frequency reflections onlap the bedrock surface, which is overlain by up to 52 m of unconsolidated sediments.

reflection-free facies lies a series of laterally continuous, high-frequency reflections that onlap an undulatory bedrock surface. Kantrowitz (1970) reported that well M-17 (Fig. 5) was drilled to a subsurface depth of 29 m, with the lowermost 24 m composed of fine-grained sands presumably equivalent to the transparent or reflection-free facies observed on the seismic sections. Laterally continuous, parallel reflections overlie bedrock along the westernmost portion of line M01.

Seismic data along line M03 (Fig. 9) suggests a similar irregular bedrock surface that projects to a maximum depth of 47 m. Reflections within the sediment-fill are coherent and laterally continuous in the western portion of this seismic profile but become more chaotic and less continuous to the east. This lateral change in acoustic facies suggests a transition from finer grained sediments beneath the wetlands to coarser grained deposits under drumlinized topography (Fig. 3).

Maximum depth to bedrock is 52 m in a trough along the east-central portion of seismic line M01 (Fig. 9). This depth is in good agreement with well M-36, located 3 km south of line M01, that bottomed in sediment at a depth of 56 m (Fig. 5). The seismic reflection data from central Montezuma suggest erosion to subsurface depths greater than at the southern end of the wetlands (seismic lines M08 and M10; gas wells M-1a and M-1b).

Seismic lines M05 and M06. Two seismic reflection profiles (M05 and M06, Fig. 10), totaling a distance of 2.5 km, were obtained 16 km northwest of Cayuga Lake (Fig. 3). These seismic lines traverse wetland areas and are separated by approximately 1.5 km of drumlinized (higher elevation) topography (Fig. 3). Lines M05 and M06 reveal an irregular, high-amplitude bedrock reflection that ranges between subsurface depths of 20 to 32 m (Fig. 10). Bedrock depths are greatest along the center of the seismic lines with a gradual shallowing along the margins (Fig. 10). A thicker, transparent reflection facies overlies bedrock within the wetland areas (Fig. 10), whereas chaotic reflec-

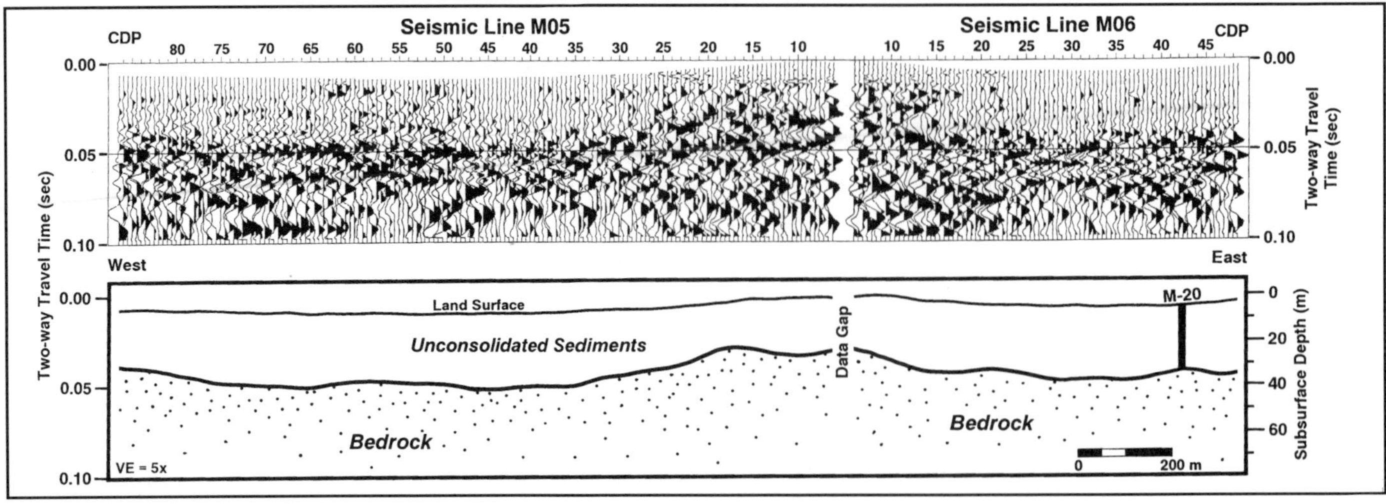

Figure 10.Weight-drop, multichannel seismic reflection profile M05 and M06 (top), and line drawing interpretation (bottom) collected across the western branch of Montezuma channels. Up to 32 m of unconsolidated sediment overlies the bedrock surface. Bedrock shallows in areas of drumlinized topography.

tions cap bedrock beneath drumlinized areas. Well M-20, located near the eastern end of seismic line M06 (Fig. 3), encountered bedrock at a subsurface depth of 27 m (Fig. 5). Bedrock here is overlain by 14 m of fine-grained sand that correlates with the reflection-free seismic facies recorded on line M06 (see Fig. 10).

Northern Montezuma wetlands

Seismic line M07. Seismic reflection line M07 (0.6 km, Fig. 11) traversed a narrow channel approximately 21 km north of Cayuga Lake, in the northwestern portion of the Montezuma wetlands (Fig. 3). Line M07 is characterized by an irregular, high-amplitude bedrock reflection overlain by more than 0.03-sec TWT of a transparent acoustic facies (20 to 35 m thick, Fig. 11). This transparent facies thins westward where a drumlin is cored by discontinuous reflections. Bedrock here reaches a maximum depth of 38 m. Well M-29, located approximately 2 km west of M07 (Fig. 3), beyond the limits of the wetlands, penetrated sand and clay to a subsurface depth of 5 m, then encountered coarse-grained sand from 5 to 8 m, and finally 21 m of glacial till or "hardpan" (Fig. 5). Approximately 1.5 km east of line M07, well M-32 (Fig. 5) penetrated 32 m of unconsolidated sediments before encountering bedrock.

Seismic line M09. Seismic reflection line M09 (~1 km) traversed the northeastern end of Montezuma wetlands across its eastern branch approximately 21 km north of Cayuga Lake (Fig. 3). Chaotic reflections near the center of the profile overlie an irregular, high-amplitude bedrock reflection (Fig. 12). The channelized surface beneath the wetlands portion of the profile reveals a maximum depth to bedrock of 38 m. However, bedrock shallows along the drumlinized western margin of line M09 to a subsurface depth of 20 m (Fig. 12).

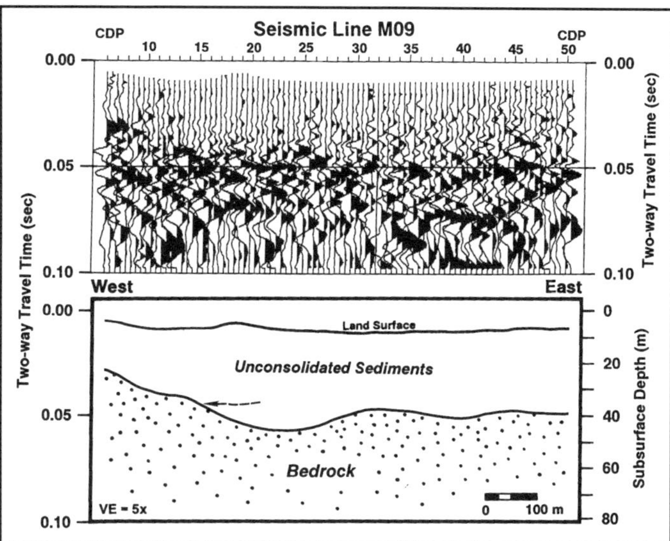

Figure 12. Weight-drop, multichannel seismic reflection profile M09 (top) and line drawing interpretation (bottom) acquired at the northern end of the Montezuma channels' eastern branch. The bedrock surface appears to shallow beneath a drumlin along the western portion of the seismic line. Maximum depth to bedrock is 38 m.

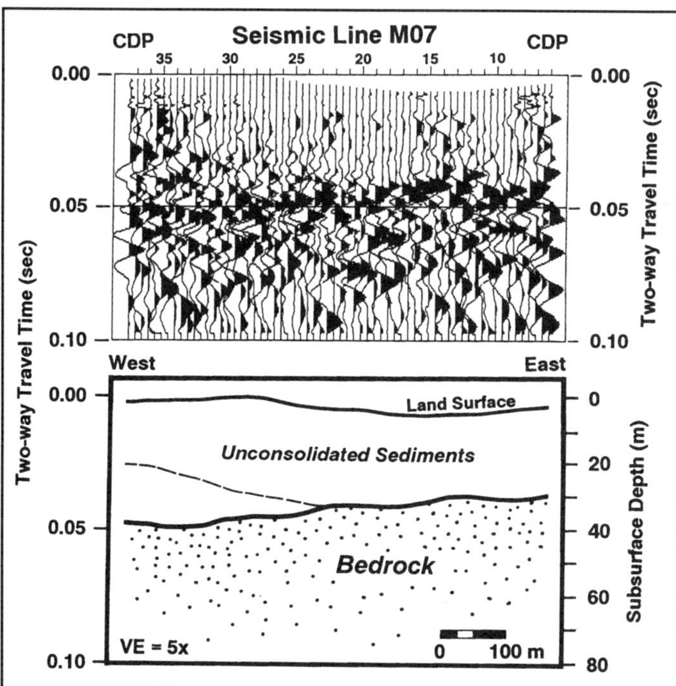

Figure 11. Weight-drop, multichannel seismic reflection profile M07 (top), and line drawing interpretation (bottom) acquired at the northern end of the Montezuma channels' western branch. Maximum depth to bedrock is 38 m.

Bedrock morphology

On the basis of the seismic reflection data presented in this study that were correlated with available borehole records (unpublished U.S. Geological Survey database, Ithaca, New York; Cox and Lewis, 1965; Kantrowitz, 1970), four east-west cross sections were constructed to illustrate the bedrock morphology and thickness of sediment-fill beneath the Montezuma wetlands (Fig. 13).

The northernmost cross section (A-A′) traverses the wetlands approximately 21 km north of Cayuga Lake (Fig. 3). Although well data are scarce from this area, seismic reflection data indicate that bedrock reaches a maximum subsurface depth of only 38 m below the wetland surface. Well M-32, which was drilled on an interchannel bedrock high, indicates bedrock at a depth of 32 m.

Cross section (B-B′) located approximately 15 km north of Cayuga Lake (Fig. 3), depicts a nearly symmetrical bedrock channel (Fig. 13) along the eastern branch of the wetlands where maximum sediment thickness is at least 50 m. The western portion of cross-section B-B′ also exhibits a channelized bedrock morphology with bedrock highs at the cores of drumlins. The western channel is both narrower and not as deeply incised as its eastern counterpart (Fig. 13).

Approximately 9 km north of Cayuga Lake, cross-section C-C′ traverses both the eastern and western branches of Montezuma wetlands (Fig. 3). Separating the wetland areas is a

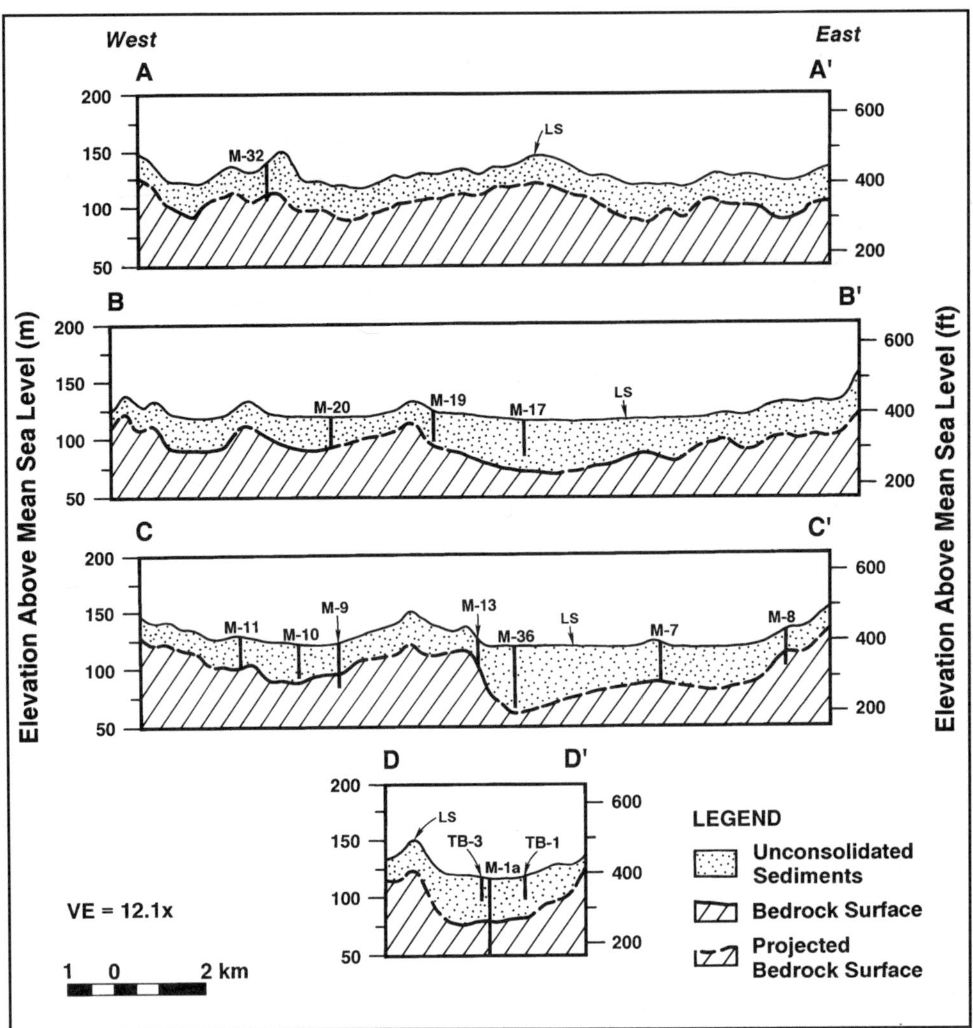

Figure 13. Four cross sections created from seismic reflection data as well as water-well, gas-well, and test boring records. Bedrock is asymmetrical in cross section and undulates from north to south. Locations of cross sections are shown in Figure 3. LS = land surface.

drumlinized topographic high. This cross section depicts an undulating bedrock surface that shallows beneath the drumlins. An asymmetric-shaped bedrock channel characterizes the eastern branch of the wetlands. Both wells M-13 and M-8, which border the wetlands, terminate in bedrock and indicate that bedrock shallows to the west and to the east. Well M-14, located 1.5 km north of profile C-C′ (Fig. 3), bottomed in gravel at a subsurface depth of 39 m (Fig. 5). Well M-11 similarly penetrated sand and gravel at a subsurface depth of 30 m (Fig. 5). The Montezuma wetlands' western branch in this area is characterized by shallower depths to bedrock (<40 m).

The most southerly cross-section (D-D′), approximately 1.5 km north of Cayuga Lake (Fig. 3), depicts a somewhat symmetrical, trough-shaped bedrock surface that extends as much as 46 m beneath the wetlands (Fig. 13). Depth to bedrock is supported by gas wells M-1a and M-1b (42 m, Fig. 5). Seismic data indicate that the bedrock occurs at a shallower depth and may core the drumlins along the margin of the wetlands. U.S. Geological Survey test boring data (TB-1, TB-2, and TB-3; Fig. 5) document sediment-fill as a fining-upward sequence composed of coarse sand or gravel overlain by finer grained sand, clay or interbeds of peat and marl.

In general, bedrock beneath the eastern branch of Montezuma wetlands is deeper than its western counterpart and displays a more channel-like morphology. Sediments within the wetlands consist of a fining-upward sequence from basal gravels or sands to clay to peat/marl deposits. Drumlins are either cored by bedrock or composed of coarse-grained sediments that overlie bedrock.

DISCUSSION

The land-based, multichannel seismic reflection data of this study have been integrated with available water-well, gas-well, and test boring data to reveal the depth to bedrock and sediment thickness beneath the Montezuma wetlands. Seismic profiles document an irregularly shaped, undulatory, channelized erosion surface beneath the wetlands that is overlain by up to 56 m of fining-upward unconsolidated sediments. Based on these results and surficial geologic data (Muller and Cadwell, 1986), the southward-directed paleodrainage system of the Montezuma wetlands appears to be a glacial meltwater channel system. Two scenarios are possible for the formation of this drainage network: a proglacial fluvial system, or a subglacial meltwater system.

Formation of the channels as a proglacial meltwater channel system would require a stabilized ice margin north of the present-day Montezuma wetlands. However, Muller and Cadwell (1986) do not map any direct evidence for an ice margin within or directly north of the Montezuma channels (see Fig. 2). The only till moraines in the vicinity of the Montezuma channels occur at their *southern* margin (Muller and Cadwell, 1986) (Fig. 2). The Montezuma channel system is also unusually broad, displaying a maximum width of 6 km over a north-south distance of only 25 km. Furthermore, the undulatory nature of the bedrock surface and associated bedrock sill near the northern end of Cayuga Lake could not be the result of overland (subaerial) stream flow. Also, erosion of the channels by subaerial streams to depths of ~50 m below the modern level of Cayuga Lake would require a lowering of local base level (i.e., lake level) by at least 50 m for which there is no evidence. Thus it does not appear as though the Montezuma channels were cut by a subaerial, south-flowing proglacial stream system.

An alternate hypothesis is that the Montezuma channels were eroded by subglacial meltwaters that were focused into this topographic low of the Finger Lakes region. The paleodrainage pattern associated with the wetlands today (see Figs. 1 and 2) is similar in scale and pattern to subglacial meltwater channels described from the Puget Lowlands of Washington State (Booth and Hallet, 1993) and south-central Ontario, Canada (Brennard and Shaw, 1994). A subglacial origin for Montezuma channels would also allow a direct evolution of facies from water-laid, coarse-grained sediments (subglacial fluvial) to finer grained sediments (glaciolacustrine and lacustrine) without an intervening subaerial stage during ice sheet recession.

Unlike gravity-driven, subaerial streams, subglacial meltwater flow is a pressure-driven system that can drain large areas (volumes) of ice sheets (Röthlisberger, 1972; Shreve, 1972). Pressurized subglacial meltwater flow could account for the undulatory nature of the bedrock surface beneath the Montezuma channels. In fact, "humped" or "up and down" longitudinal profiles constitute a common criteria for channels eroded by subglacial meltwater (Booth and Hallet, 1993). Pressurized subglacial meltwater flow could also explain how the Montezuma channels were eroded ~50 m below local base level. Tunnel valleys, which are eroded by subglacial meltwater, are known to extend hundreds of meters below sea level (Boyd et al., 1988). Subglacial meltwater drainage of a large ice sheet like the Laurentide, could also provide the flow volume necessary to erode a channel system like Montezuma that is shallow (tens of meters) but up to 6 km wide extending over a length of only 25 km.

A block diagram reconstruction of the Montezuma wetlands (Fig. 14A) depicts this hypothesis of a subglacial meltwater drainage system during the late Wisconsin. Erosion was restricted to a series of channels with deeper excavation in areas that are underlain by finer grained, clastic sedimentary rocks of the Silurian Salina Group. The Onondaga Limestone acts as a cap rock to weaker Salina Group shales (Coates, 1968; Rickard and Fisher, 1970). The bedrock sill near the north end of Cayuga Lake consists of the more resistant Onondaga Limestone.

If the Montezuma channels were eroded by subglacial meltwaters, the eroded material should have been deposited in Cayuga Lake as proglacial sediment. A northward thickening wedge (~55 m) of proglacial to glaciolacustrine sediments (Mullins et al., 1991) is in fact present near the northern end of Cayuga Lake. These sediments appear to have been transported from a proximal source area (Montezuma) and are stratigraphically younger than the Valley Heads Moraine deposits south of the Finger Lakes (Mullins and Hinchey, 1989). Radiocarbon dates constrain the age of the Valley Heads Moraine and subsequent sediment infill of the Finger Lakes to between 14.4 and 13.6 ka (Mullins and Hinchey, 1989; Mullins et al., 1991). Thus, excavation of the Montezuma channels (Fig. 14A) and deposition of the sediment wedge in Cayuga Lake can be constrained to sometime between ~14.4 and 13.6 ka, and most likely occurring toward the younger end of this range.

Basal channel-fill sediments in Montezuma were probably deposited under subglacial fluvial processes when the ice terminus was situated near the north end of present-day Cayuga Lake (till moraines; see Fig. 2). Drumlinization at the north end of Cayuga Lake may have occurred during a phase of active meltwater production (White, 1985) contemporaneous with subglacial deposition within channelized low areas (Fig. 14B).

Ultimate ice sheet withdrawal from the northern Cayuga Lake basin (Fig. 14C) resulted in a drainage reversal by allowing a northern outlet for highstanding proglacial lake waters (Mullins and Hinchey, 1989). Figure 14D depicts conditions during this time when proglacial lake levels fell due to northward recession of the Laurentide ice sheet from the Finger Lake basins. Postglacial lacustrine sediment accumulated within low-lying areas of the Montezuma channels during the latest Pleistocene-Holocene as regional lake levels fluctuated in response to climatic/hydrologic changes.

Sediment cores recovered from Crusoe Lake within Montezuma wetlands (M-35, Figs. 3 and 5) record the nature of this postglacial sediment infill. Timing of sediment deposition beneath Crusoe Lake (Cox and Lewis, 1965) is based on two radiocarbon dates of 6,850 ± 150 B.P. at 5-m subsurface and 3,200 ± 80 B.P. at 1.3-m subsurface. Sediment in these cores consists of marly, gyttja-containing shell fragments, indicating that postglacial lacustrine deposition occurred.

Today, underfit, meandering streams flow northward out of Cayuga Lake through Montezuma wetlands (Fig. 14E) with

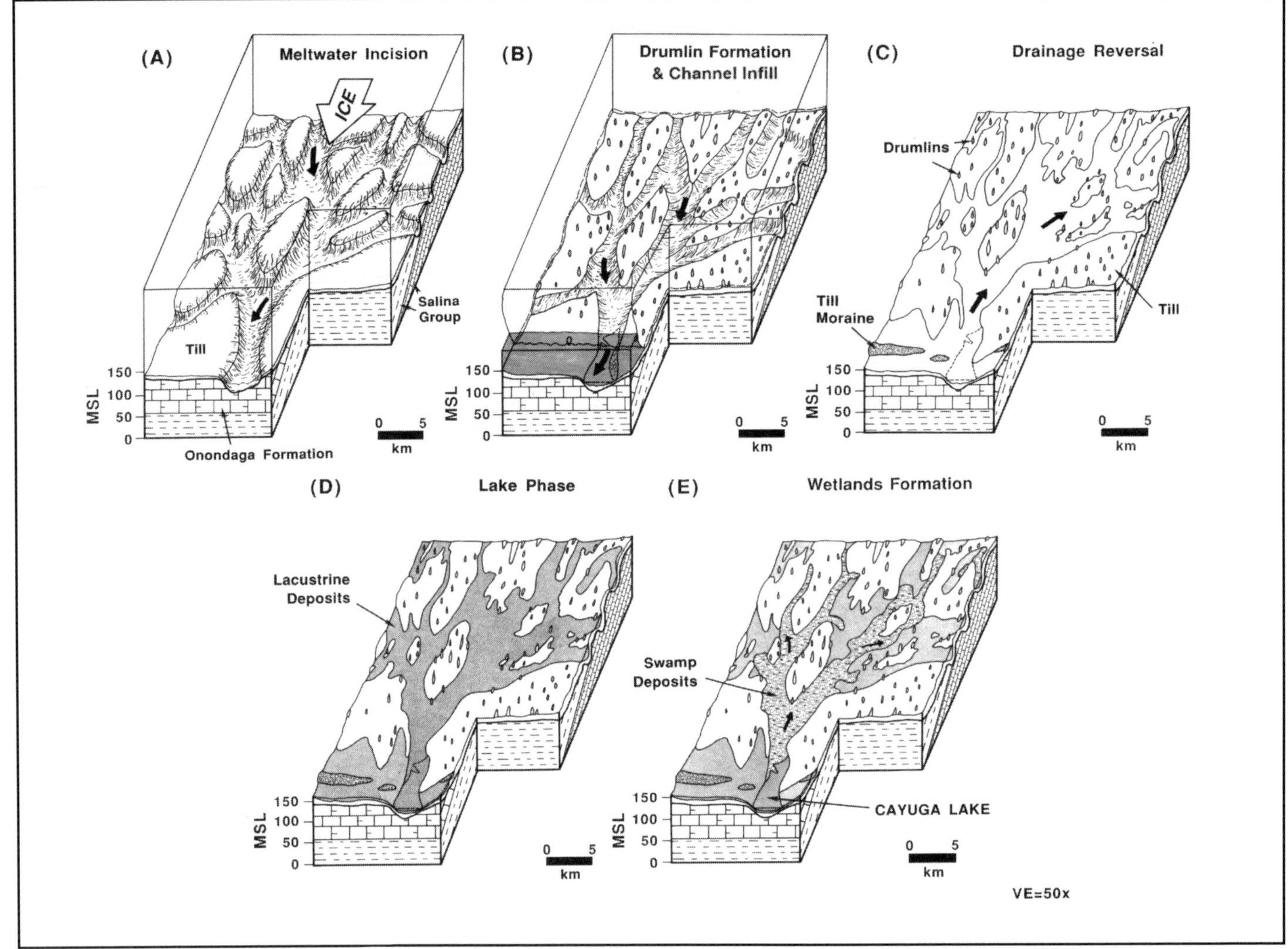

Figure 14. Schematic block diagram depicting the geomorphic evolution of Montezuma channels. A, Sediment and bedrock incision by southward-flowing channelized subglacial meltwater; B, drumlin formation and channel infill; C, ice recession and drainage reversal to the north; D, lacustrine sedimentation during post-glacial lake phase; E, contemporary wetland development in low-lying areas. MSL = mean sea level (m).

similar conditions having existed for the last ~7,000 years (Cox and Lewis, 1965). During the past two centuries, increased drainage and runoff of the wetlands have occurred due to human intervention, further reducing the size and lateral extent of the wetlands (Swerdlow, 1990).

SUMMARY AND CONCLUSIONS

This study has provided a shallow subsurface evaluation of the relict drainage pattern presently displayed by the Montezuma wetlands. Our results indicate that:

1. The erosion surface extends at least 50 m below the local base level of modern Cayuga Lake.

2. The erosion surface consists mainly of undulatory, asymmetrical channels that slope southward before rising over a bedrock sill near the north end of Cayuga Lake.

3. Maximum sediment thickness beneath Montezuma wetlands is at least 56 m.

4. Sediment thickness and depth to bedrock are greater beneath wetland and remnant lacustrine areas than beneath interchannel, drumlinized areas.

5. Borehole data indicate a fining-upward sequence of coarse to fine sediments beneath channel areas, whereas only coarser grained, poorly sorted sediments mantle interchannel bedrock highs.

Based on the nature of the bedrock morphology and sediment-infill of the Montezuma wetlands, the following late Quaternary evolutionary scenario is proposed:

1. Pressurized subglacial meltwater flow at the base of the southern margin of the Laurentide ice sheet resulted in channel

excavation during regional deglaciation sometime between ~14.4 and 13.6 ka.

2. Drumlin formation on interchannel areas within Montezuma wetlands occurred contemporaneously with meltwater channel incision or during coarse-grained, basal channel infill.

3. Retreat of the Laurentide ice sheet from the northern Cayuga Lake basin ~13.6 ka resulted in a regional drainage reversal.

4. Postglacial lakes occupied the Montezuma channels and resulted in fine-grained lacustrine sediment deposition.

5. Modern wetland conditions developed sometime after 7 ka, once shallow postglacial lakes drained.

ACKNOWLEDGMENTS

This research was supported by National Science Foundation Grants EAR–8607326 and EAR–8903870 awarded to H. T. Mullins. Special thanks to H. T. Mullins for guidance and critical reviews. Additional thanks to the staff of the Department of Earth Sciences, Syracuse University; and to R. S. Darling, C. Padover, C. B. Andersen, and B. T. Aulenbach for field work assistance. W. M. Kappel and W.S. McPherson, U.S. Geological Survey, Ithaca, New York, provided well logs. I also thank the staff of Montezuma National Wildlife Refuge for permission to conduct this project. Reviews by P. U. Clark, Oregon State University, and R. LeB. Hooke, University of Minnesota, were of great value in revising this chapter.

REFERENCES CITED

Bloom, A. L., 1986, Geomorphology of the Cayuga Lake basin, *in* New York State Geological Association, 58th Annual Meeting Field Trip Guidebook: Ithaca, New York, Cornell University, p. 201–279.

Booth, D. B., and Hallet, B., 1993, Channel networks carved by subglacial water: Observations and reconstruction in the eastern Puget Lowland of Washington: Geological Society of America Bulletin, v. 105, p. 671–683.

Boyd, R., Scott, D. B., and Douma, M., 1988, Glacial tunnel valleys and Quaternary history of the outer Scotian shelf: Nature, v. 333, p. 61–64.

Brennard, T. A., and Shaw, J., 1994, Tunnel channels and associated landforms, south-central Ontario: Their implications for ice-sheet hydrology: Canadian Journal of Earth Sciences, v. 31, p. 505–522.

Burger, H. R., 1989, Delineation of aquifer geometries, Connecticut Valley, Massachusetts, using geophysical techniques: Second Keck Research Symposia in Geology, Colorado Springs, Colorado, p. 126–129.

Coates, D. R., 1968, Finger Lakes, *in* Fairbridge, R. W., ed., Encyclopedia of geomorphology: New York, Reinhold Books, p. 351–357.

Coates, D. R., 1974, Reappraisal of the glaciated Appalachian Plateau, *in* Coates, D. R., ed., Glacial geomorphology, Publications in Geomorphology: Binghamton, State University of New York, p. 205–243.

Cox, D. D., and Lewis, D. M., 1965, Pollen studies in the Crusoe Lake area of prehistoric Indian occupation: New York State Museum and Science Service, Albany, Bulletin, no. 397, p. 1–29.

Dix, C. H., 1955, Seismic velocities from surface measurements: Geophysics, v. 20, p. 68–86.

Dobrin, M. B., and Savit, C. H., 1988, Introduction to geophysical prospecting (fourth edition): New York, McGraw-Hill, 867 p.

Fairchild, H. L., 1907, Drumlins of central and western New York: New York State Museum and Science Service, Albany, Bulletin, no. 111, p. 391–443.

Fairchild, H. L., 1934, Cayuga Valley lake history: Geological Society of America Bulletin, v. 45, p. 233–280.

Hodgson, P. R., 1984, Application of microcomputer processing of shallow seismic reflection data [M.S. thesis]: West Lafayette, Indiana, Purdue University, 70 p.

Hunter, J. A., Pullan, S. E., Burns, R. A., and Good, R. L., 1984, Shallow seismic reflection mapping of the overburden-bedrock interface with the engineering seismograph. Some simple techniques: Geophysics, v. 49, p. 1381–1385.

Kantrowitz, I. H., 1970, Ground-water resources in the Eastern Oswego River basin, New York: Albany, New York State Conservation Department, Division of Water Resources, 129 p.

Li, H. C., 1991, Cretaceous and Tertiary seismic stratigraphy of the Coastal Plain of southern New Jersey [Ph.D. dissertation]: New Brunswick, New Jersey, Rutgers University, 289 p.

Miller, R. D., Pullan, S. E., Waldner, J. S., and Haeni, F. P., 1986, Field comparison of shallow seismic sources: Geophysics, v. 51, p. 2067–2092.

Miller, R. D., Steeples, D. W., and Myers, P. B., 1990, Shallow seismic reflection survey across the Meers fault, Oklahoma: Geological Society of America Bulletin, v. 102, p. 18–25.

Muller, E. H., and Cadwell, D. H., 1986, Surficial geologic map of New York—Finger Lakes sheet: Albany, New York State Museum, Geologic Survey Map and Chart Series 40, 1 sheet, 1:250,000.

Mullins, H. T., and Hinchey, E. J., 1989, Erosion and infill of New York Finger Lakes: Implications for Laurentide Ice Sheet deglaciation: Geology, v. 17, p. 622–625.

Mullins, H. T., Wellner, R. W., Petruccione, J. L., Hinchey, E. J., and Wanzer, S., 1991, Subsurface geology of the Finger Lakes region, *in* Ebert, J. R., ed., New York State Geological Association, 63rd Annual Meeting Field Trip Guidebook: Oneonta, State University of New York, p. 1–54.

Petruccione, J. L., 1994, Bedrock morphology and geomorphic evolution of Montezuma wetlands, central New York state: A subglacial meltwater discharge system beneath the southern Laurentide Ice Sheet? [M.S. thesis]: Syracuse, New York, Syracuse University, 110 p.

Rickard, L. V., and Fisher, D. W., 1970, Geologic map of New York—Finger Lakes sheet: Albany, New York State Museum, Geologic Survey Map and Chart Series 15, 1 sheet, 1:250,000.

Ridky, R. W., and Bindschadler, R. A., 1990, Reconstruction and dynamics of the late Wisconsin "Ontario" ice dome in the Finger Lakes region, New York: Geological Society of America Bulletin, v. 102, p. 1055–1064.

Röthlisberger, H., 1972, Water pressure in intra- and subglacial channels: Journal of Glaciology, v. 11, p. 177–203.

Shreve, R. L., 1972, Movement of water in glaciers: Journal of Glaciology, v. 11, p. 205–214.

Somanas, C. D., Bennett, B. C., and Chung, Y. J., 1988, In-field seismic data processing and Eavesdropper, user's manual V1.1: Lawrence, Kansas Geological Survey, Computer Program Documentation 88–4, 82 p.

Swerdlow, J. L., 1990, Erie Canal: Living link to our past: National Geographic Magazine, v. 178, no. 5, p. 38–66.

von Engeln, O. D., 1961, The Finger Lakes region: Its origin and nature: Ithaca, New York, Cornell University Press, 130 p.

White, W. A., 1985, Drumlins carved by rapid water-rich surges: Northeastern Geology, v. 7, p. 161–166.

Manuscript Accepted by the Society January 16, 1996

Printed in U.S.A.

Typeset and printed in U.S.A. by Johnson Printing, Boulder, Colorado